WORLD PERSPECTIVES

12

Matter, Mind and Man

WORLD PERSPECTIVES

Planned and Edited by Ruth Nanda Anshen

*

Jacques Maritain
Approaches to God
Walter Gropius
Scope of Total Architecture
Radhakrishnan
Recovery of Faith
Brock Chisholm
Can Nations Learn to Learn?
Lewis Mumford
The Transformations of Man
Konrad Adenauer
World Indivisible
V. G. Childe
Society and Knowledge
Fred Hoyle
Man and Materialism
Eric Fromm
The Art of Loving
Paul Tillich
Dynamics of Faith
D. T. Suzuki
Mysticism
Denis de Rougemont
Man's Western Quest

WORLD PERSPECTIVES

EDMUND W. SINNOTT

Matter, Mind and Man

Ruskin House

GEORGE ALLEN & UNWIN LTD

MUSEUM STREET LONDON

First published in Great Britain in 1958

Printed in Great Britain
in 11pt. Pilgrim type
by
East Midland Printing Company Limited
Bury St. Edmunds, Peterborough, Kettering
and elsewhere

Contents

I

The Paragon of Animals

'WHAT a piece of work is a man!' says Hamlet, 'how noble in reason! how infinite in faculty! in form and moving how express and admirable! in action how like an angel! in apprehension how like a god! the beauty of the world! the paragon of the animals!'

These familiar words express the admiration which must be ours, as well, for the remarkable species to which we all belong. Whatever our particular philosophy about him may propose, whether we think him risen ape or fallen angel, he stands unique among all living things. His entrance on the scene of the great evolutionary drama has been its crashing climax and has changed the plot competely. How this drama will continue to unfold and whether in the end he will be the hero or the villain of it we cannot yet be sure, but the centre of the stage is now completely his.

To see what his significance has been in the earth's story, one should attempt to picture him against the background of the long past from which he has emerged. If an observer from another planet could have visited our earth from time to time through the past geological eras and have watched the history of life unfolding, a history reckoned not by centuries but by ten thousand centuries multiplied a thousand fold, he would have noticed in the warm primeval seas the first faint stirrings of that strange phenomenon we now call life. How it arose is still a mystery but somewhere, somehow, protoplasm came to being and, in primordial masses or in tiny cells, began those regulated processes of metabolism, growth and reproduction that have ever since distinguished living things. With ponderous slowness these primitive creatures changed and grew to more complex ones—the simple seaweeds first and their allies, then sponges and jellyfishes, worms, mollusks and crustacea and finally the ancient fishes, first of vertebrates. Throughout this early time, some billion years perhaps, all life was in the sea or along the shores of it. In the Silurian era came the first steps in the invasion of the land. Up till now this had been uninhabited, masses of bare and barren rock, desert and lifeless. Its conquest by living creatures would have impressed our observer as one of the most dramatic events in the earth's history.

To live on land presented problems of the utmost difficulty. Primitive plants were able to establish themselves there by developing roots that could draw water from the soil, protective layers for their leaves that kept watery parts of them from drying out but allowed exchanges with the atmosphere, and other tissues that held up their bodies firmly in the air. Animals followed plants into the new environment and became adapted to take oxygen from the air. With these modifications for life on land, evolutionary progress moved much faster. Larger plants appeared with well-developed roots and stems and leaves, the forerunners of the plant life of today. Among animals came amphibians, adapted to life in water and on land. The first members of some other animal types now came to being. All this reached a magnificent crescendo in the rich and varied life of the Carboniferous era, flourishing in unheard of abundance, especially in the wide-spread estuaries and marshes. Here, our observer doubtless would have been convinced, was to be seen the climax of the evolutionary process—not only the oceans filled with living things but the land itself covered with vegetation and abundantly populated with animals of many kinds. Advance could obviously go no further.

But this was only the beginning. From simpler fern-like plants evolved the coniferous trees that made forests of the sort we know today. More amazing still, the primitive amphibia, after flourishing in the Triassic, gave place to the first true land animals, the reptiles. In a variety and profusion that impresses us even now when only their bare bones are left, those strange beasts, the dinosaurs, ran, swam and flew on the face of the earth, invading almost every habitat and showing what amazing possibilities had been developed in the animal kingdom.

The earth of those days and especially its reptile-dominated portions must have been a place of wonders and it still excites our imagination. Surely now our planetary visitor must have thought that he was watching the consummation of the evolutionary drama.

With the coming of the Tertiary, however, progress proved itself by no means at an end. The flowering plants we know so well today now came to being. With them appeared teeming hosts of insects and of birds. Reptiles lost their dominance and from an inconspicuous ancestral group there rose the mammals, warm-blooded, far more intelligent and with better means for reproduction. These prospered amazingly and evolved more varied kinds than ever the reptiles had been able to produce. The land was like a zoological garden filled with a great variety of beasts that only unspoiled tropical Africa could approach in modern times. Many of the most remarkable forms have vanished—notably such huge ones as the cave bear and the sabertooth, mammoths and elks and other giant herbivores. Even to our own eyes this richness of plant and animal life seems to be the highest point, except for man, that organic evolution could ever reach. We can hardly imagine another great step forward such as earlier geological ages had experienced.

But now came something so amazing that it completely changed the course of the earth's history. Out of an obscure and inconsiderable group of ape-like mammals rose the early primate stock. They had larger brains, opposable thumbs, and lacked an overspecialization in any one direction—speed, size, protective or aggressive armaments—that would prevent them from accommodation to activity of many kinds. One of the members of this group—just how he was related to the rest we do not know—learned to stand on his hind feet and began to develop towards something like a man. For hundreds of centuries he was little more than another animal and had no great effect upon the rest or on the face of nature. But as time went on, the important way in which he differed from the other beasts—his larger brain—began to have great consequence, for he slowly became able to do what they could not—to think and reason.

The ascent of man to his present high estate has been a long and difficult journey. As to many of its steps we still are ignorant, but the main course can now be plotted with some assurance. Our remote ancesters were doubtless ape-like and probably tree-dwellers who swung from branch to branch in the comparative safety of an arboreal life. Some twenty million years ago or thereabouts, a more enterprising member of that stock descended from the trees to the ground. Instead of walking on all fours as did many apes, he chiefly used his legs and feet for locomotion and thus obtained the immense advantage of freeing arms and hands for other uses. His opposable thumb could grasp objects readily, an ability of much importance when his descendants began to handle tools. As to just what these creatures were, we are yet uncertain. *Dryopithecus* may have been one, or *Proconsul* from East Africa, both in the Miocene period. Sometime later, probably, came *Australopithecus,* the South African ape-man, a slightly higher type.

From these ancient and partially erect ground apes, by long, slow changes lasting millions of years, rose animals with larger brains and better adaptations. When it was that one of these attained a level where he might truly have been called a man is quite uncertain, and doubtless for long ages there were semi-human types which, if alive today, would seriously confuse zoologists and anthropologists. About a million years ago, however, there appeared a heavy-browed, chinless creature, *Pithecanthropus erectus,* the Java ape-man, who had a brain case with twice the capacity of our apes today and about two-thirds as large as that of man. A little later came the Peking man, still larger of brain and with a higher brow. More important as a sign of his advance is the fact that he apparently had discovered the use of fire.

Much further down the years, somewhere on the great Eurasian land-mass there arose the first creatures who were clearly human, the men of Neanderthal. Their remains are found over much of Europe, and it is they who were the 'cavemen' who so fascinate us today. They were enough unlike ourselves so that anthropologists have used a different name for them, *Homo neanderthalensis* instead of *Homo sapiens,* but men they certainly were. Their brains were as large as ours but less well developed. Their bodies were stocky and powerful and probably covered with hair. They had considerable skill in making and using tools and weapons. The more recent of them buried their dead with care and thus seem to have possessed the beginnings of a religion. In southern Africa and the east Indian region fossils of somewhat similar primitive human types have now been found.

How we are related to Neanderthal man is not yet clear, but the first true men had probably appeared before he became extinct and may have hastened his passing. The Cro-Magnon men of southern France and the Basque country were much like us, in body and in brain. They were not only highly intelligent but possessed of remarkable artistic skill, as their ancient caves now vividly make clear. From somewhere near them stems the course of our ascent.

Homo sapiens was able, as his predecessors had not been, to see the relationships of things, to perceive general facts and refer particular cases to them, and thus to reason. His memory was greatly extended and he possessed the priceless gift of imagination. Physically he was not greatly different from his primitive cousins but he had passed that critical point in evolution where intelligence emerges from simple brute mentality and takes over the course of evolution.

> Man in the rudest state in which he now exists [says Charles Darwin] is the most dominant animal that has ever appeared on this earth. He has spread more widely than any other highly organized form, and all others have yielded before him. He manifestly owes this immense superiority to his intellectual faculties, to his social habits, which lead him to aid and defend his fellows, and to his corporeal structure. The supreme importance of these characters has been proved by the final arbitrament of the battle for life. Through his powers of intellect, articulate language has been evolved, and on this his wonderful advancement has mainly depended. . . . He has invented and is able to use various weapons, tools, traps, etc., with which he defends himself, kills or catches prey, and otherwise obtains food. . . . He has discovered the art of making fire, by which hard and stringy roots can be rendered

> digestible, and poisonous roots or herbs innocuous. This discovery of fire, probably the greatest ever made by man, excepting language, dates from before the dawn of history. These several inventions, by which man in the rudest state has become so pre-eminent, are the direct results of the development of his powers of observation, memory, curiosity, imagination and reason.[1]

Physically man changed little after this, but his life greatly altered. He drew together into communities, domesticated animals, began to practise agriculture, invented the bow and the wheel and learned the use of metals. Finally, about ten thousand years ago, the earliest form of what we call civilization came into being.

The changes since that day have been tremendous. Man's advances in technical knowledge grew steadily and since the advent of science have been fantastic. He has gained control of vast sources of energy. He has learned to combat disease. He can minister to his comfort and convenience in endless ways. He has, so to speak, taken the evolutionary process into his own hands; no longer a physical and bodily evolution but a social and cultural one. Intelligent guidance, not mere survival value, is now the cause of change. Under his impact the very face of the earth has radically altered. Forests have gone, cities have risen, roads have been constructed. Even the air is filled with his machines. No other species has ever had anything like the effect on its surroundings that this naked, pushful, reasoning biped has produced.

Our planetary observer would have been startled at the sudden, almost explosive changes that man made in the earth. This human period has lasted but a moment in terms of the long, slow progress of evolution before man came. Furthermore, during this time the pace of change has steadily been increasing and it now is headlong. With the entrance of man the drama of life seems to be coming to a sudden climax. What he will do or be in the next few centuries is more than the liveliest imagination now can picture. Perhaps a new act is ready to begin. Perhaps at last the curtain will ring down.

[1] Charles Darwin, *The Descent of Man,* Ch. II.

However this may be, the most important animal on earth, or so it has always seemed to him, is man. With sublime self-assurance and the support of Holy Writ, he has assumed that all created things have been placed under his dominion and exist for his especial benefit. The beasts of the field, the fowls of the air and the fish that swim in the sea all are his. Herbs are for the service of man, as are the trees of the forest and the lilies of the field. This exalted position was a natural one for him to take, for is he not obviously wiser and stronger than all these, and thus of greater consequence? Surely he is of more value than many sparrows. Unworthy he may sometimes be, but never worthless. The religion of humanism regards him as such a superior creature that it has put man at the centre of its reverence in place of God.

Man's greatness lies not only in the power of his mind. Within him there are qualities that beasts do not possess. He has discovered beauty in nature, delighted in it and interpreted it in his art. He has pondered the deep questions of philosophy. He has framed moral codes to regulate his conduct and in him is an inner monitor to help him to distinguish right from wrong. He has been moved with awe and reverence in the presence of what seems to him Divine, and aspires to hold communion with it. Such qualities lift him up above all other creatures.

This unique place man holds has always excited his interest in his own origin and how he came to be. Every religion has its myths about this great event. Whether it is Deucalion and Pyrrha tossing stones behind them or one of many others, man's creation always is portrayed as different from that of the beasts. The finest of these stories is the one in the first chapter of Genesis, where his birth is the climax of the whole creative process and brings it to a close. Here God fashions man from the dust of the ground but in His own image, and breathes into him a living soul.

This belief in a spiritual side of man, in the presence in him of something more than his physical body, occurs in the ideas he has always held about himself. He seems to possess a psyche, a soul, a spirit, something immaterial that guides his acts and is often thought of as distinct from his body and even capable of surviving death itself. The origin of this belief about the soul goes deep into human history and stems from man's early experiences. Even the cave-men would not have buried their dead so carefully unless they had thought that a man was something more than his body. Dreams, too, doubtless had an important share in this belief, for primitive man often fails to distinguish between these and real experiences. The spirit world seemed to surround early man. So many strange happenings had no evident basis in physical causes that he assumed spirits were everywhere at work and thus came to placate and finally to worship them. 'Some explanation of the phenomena of life,' says Darwin, quoting from one of his contemporaries, 'a man must feign for himself; and to judge from the universality of it, the simplest hypothesis, and the first to occur to men, seems to have been that natural phenomena are ascribable to the presence in animals, plants, and things, and in the forces of nature, of such spirits prompting to action as men are conscious they themselves possess.'[1] His own spirit was obviously related, so he thought, to those spirits around him, and through it he tried to establish a communion with all nature.

As religion advanced, these concepts became more refined. Christianity maintains man's unique position and the supreme value of his soul. The importance of saving it from perdition has provided much of his motivation, and the possibility of its communion with God and of its immortality have been for centuries the high hope of Christians. Man cannot complete himself, religion says, until he finds his spiritual place in nature.

All this is part of the ancient philosophical belief that man is a double being, a body *and* soul, and that these two parts of him, though of necessity intimately related, are distinct and different things. The body is material; something, many believe, to be repressed and outgrown. The soul is spiritual and eternal, the promise of a far greater life. What sets man off from all the rest of nature, says the traditional view of him, is that he alone possesses this spiritual part, that he alone can share in the Divine.

For generations, therefore, man has thought of himself as that for which the long laborious ages have prepared; a being

[1] *Ibid.*, Ch. III.

set apart and significant in the universe; lifted by his reason and his spirit above all created things; a little lower than the angels, to be sure, but alight with a touch of the Divine flame and clothed with glory and honour. Created from the dust though he might be, he still believed himself a child of God and precious in the divine scheme of things.

This exalted opinion of his position has not been simply an expression of his vanity and egotism. It was vitally necessary. From it he drew courage and comfort in times of tribulation. Without this conviction of his worth and this belief in his relationship to the Divine, it is doubtful whether man would have been able to survive the ordeals of his eventful history. A sense of his own significance has constantly reassured him.

The crisis of our day comes from the fact that this traditional appraisal of man's nature is now gravely challenged, and from several directions. Whether in the light of modern knowledge he can maintain this high estimate of himself or must give it up for something very different, something far less exalted and godlike, is the deepest question that he has to face. Here the great traditions of the past come most sharply into conflict with man's new scientific insights about himself and the universe. How this conflict will be decided and what man's true nature will finally prove to be are questions that shake the world today. It is not the change in our ideas about God that is so ominous, important as this is, but the change in what we think about ourselves, for this will finally be reflected in our philosophy and our religion.

Is man simply the paragon of animals or is he something more? Everywhere he is trying, as never in the past, to find what manner of creature he really is. In this search all other questions have a part. Here finally converge the disciplines of science. Here come to focus the problems of philosophy. Here is an essential element in every religion. Until man comes to know *himself*, all other knowledge that he gains is incomplete.

2

The Challenge

NO ONE can doubt man's unique place in nature, his supreme eminence among the animals or the great possibilities that lie before him. As the result of modern scientific knowledge, however, the still more exalted position he once seemed to hold has been faced with such a serious challenge that the traditional ideas about him have now lost much of their old authority. The concepts of his soul, his spiritual nature, his divine origin and his hope of life beyond the grave have been relegated by many thoughtful people to the category of parable, myth and the illusion of ancient superstition. Even his mind is thought by some to be nothing but an unreal thing, a convenient name for his behaviour and little more. This change in fundamental belief is of far more than academic consequence to us today, for if the challenge it presents can be sustained, man's view of his own nature will be so radically modified as to alter his whole philosophy of life.

Such a challenge, though never so serious as it is today, is not a new one. A minority has always rejected the authority of religious orthodoxy and held views about God and man radically different from traditional ones. They were often branded as children of perdition, though many we would call religious liberals today. In the French Revolution, however, still more radical ideas prevailed and in one great nation traditional Christianity and its attitude toward God and man was officially abandoned. Reaction from such views, especially in England and America, led to a vigorous revival of religious orthodoxy in the reign of Queen Victoria. In our Western culture faith seemed certain and secure in the early years of the nineteenth century.

Into this complacent philosophy, Darwin's great book, *The*

Origin of Species, in 1859 burst with a shattering impact. Evolution had often been discussed and generally dismissed as impossible, but in the theory of natural selection it was for the first time made not only convincing but plausible, for the means by which it had come to pass were now made clear. Almost at once great numbers of intelligent people were persuaded of the truth of the new theory, hard though it was to fit this into the rest of their beliefs.

The violent opposition to it which arose in many quarters was not so much against the new conception of how living things had changed throughout the ages; this could be reconciled with the Creation story in the Bible if one did not insist on its literal interpretation. The challenge was recognized to be far deeper than that, for it was evident at once to even the most superficial reader that if animals and plants had evolved from lower forms, man must have done so, too. In the memorable debate between the Bishop of Oxford and Professor Huxley little time was spent on lower species. Man's origin from the apes became at once the issue.

Even this radical departure from tradition might perhaps have been accepted if it could have been shown that in some way the evolutionary process involved the working out of a plan or pre-visioned programme which led inevitably to the origin of man; if what biologists today call orthogenesis had been operative. But Darwin insisted, and later evolutionists have generally supported him, that the variations which were the raw material of evolutionary progress were random ones and arose by chance and not in any sense by purposive design. This was the fatal blow, for it seemed to remove any possibility that man had been deliberately created or was a child of God and made in His own image. If chance and not divine purpose had formed man, he was indeed cast down from his exalted station, a puppet of fate and nothing more.

The interest in evolution as it applied to man became so intense that in 1871 Darwin brought together his data on the subject and in *The Descent of Man* summarized the evidence that man's origin indeed had been through the same process that was responsible for the origin of all other living things. He reproved those who confidently asserted that man's evolutionary history never can be known. 'Ignorance,' he said, 'more frequently begets confidence than does knowledge. It is those who know little, and not those who know much, who so positively assert that this or that problem will never be solved by science.'[1] These words deserve to be pondered well today.

Relentlessly Darwin marshalled arguments for the descent of man from 'some lower form.' Man bears in his bodily structure clear evidence of his descent, for it is built on the common mammalian type and is very close to that of the higher apes. If he arose by a special creative act instead of by gradual change, this similarity, extending to the most minute details, is very difficult to explain. If he is really unique in creation and made in God's image, man should be set apart from other organisms, one would think, by unmistakeable bodily differences.

Darwin pointed out that physiologically man is also like the animals. He can contract many of the same diseases from which they suffer. Some apes have a strong taste for spirituous liquors and suffer the same effects from their use that man does. One American monkey, he relates, 'after getting drunk on brandy, would never touch it again and thus was wiser than many men.'

Man's embryonic history is much like that of the other vertebrates and in its very early stages can hardly be distinguished from them. In his embryology, says Huxley, man is far nearer to the apes than the apes are to the dog. He possesses rudimentary and vestigial organs, such as the power of moving his scalp or his ears, 'wisdom teeth,' an embryonic tail and other organs, understandable only in the assumption that they have persisted from a time when his ancestors found them useful.

All these traits, it may be objected, are bodily ones; but is not man's *mind* unique? It certainly is enormously superior to those of animals, Darwin admits, but this difference is one of degree and not of kind. The beginnings of all man's psychical traits can be found among the beasts, especially the higher mammals. His instincts, though few in number, are like theirs—self-preservation, sex and mother love among them. Intelligence is not his alone, for countless examples of simple forms of it may

[1] Charles Darwin, *The Descent of Man*, Introduction.

be found in lower animals. Even the ability to reason seems to have its rudiments there and instances of it have been reported many times by Darwin and others. Though language in the human sense is lacking in them, many animals communicate by particular sounds and signals; and domestic ones certainly understand much that we say to them, some dogs being able to recognize a hundred different words or more.

Animals possess memory, sometimes extending over an interval of several years. Even imagination seems to be theirs, for they can dream. Perhaps, says Darwin, a dog baying in the moonlight may be stirred by a sensation that he only imagines. The various emotions are well displayed by animals. That fear, courage, happiness, suspicion, curiosity, wonder, vengefulness and many more are felt by them we can hardly doubt as we watch their behaviour.

Again, one may object, even these qualities are not really what is most distinctive about man. His aesthetic and moral senses, his spiritual nature, are the ones that raise him above the level of the brutes. Darwin replies by describing the taste of female birds for the bright colours and graceful plumage and songs displayed by males who seek their favours. He cites the decoration of their playing places made by bower birds. This indeed is not a highly cultivated appreciation of beauty, but surely the beginnings of it are here.

As to moral sense, he cites the strong social instincts of certain animals, their mutual helpfulness and the many cases known where an individual has sacrificed himself for the welfare of the group to which he belongs. Obedience and fidelity distinguish the behaviour of many beasts. Agassiz believed that dogs possess something very like a conscience.

As to love, surely the love of a mother for her offspring is too well known to require comment, but this noblest of feelings, Darwin continues, goes far beyond such expression. The devotion of dogs to persons has long been celebrated. 'A dog is the only thing on this earth that loves you more than he loves himself.' This surely is the beginning of a quality that man regards as one of his own highest attributes.

Even religion may have deeper roots than man. 'The feeling of religious devotion . . . consists of love, complete submission to an exalted and mysterious superior, a strong sense of dependence, fear, reverence, gratitude, hope for the future, and perhaps other elements. . . . We see some distant approach to this state of mind in the deep love of a dog for his master, associated with complete submission, some fear, and perhaps other feelings.'[1] It is perhaps not too much to maintain that a dog looks on his master as a god.

These arguments of Darwin's for the fundamental similarity of man to the higher animals have been multiplied many times since *The Descent of Man* was written. They can be explained on the assumption of man's animal origin but are very difficult to account for on the traditional theory of his special creation. Most thoughtful people now agree with Darwin's conclusions, and even Roman Catholic theologians are inclined to admit the evolution of man's body, reserving only the creation of his soul as a special act of God.

We who have always been familiar with this new idea about man's origin can hardly realize the shock it gave to the religious life of the nineteenth century. The sudden and rather dramatic character of the idea, coming as it did through the publication of a single book, was known and talked about by almost everyone. *The Origin of Species* soon became a best seller and even those who did not read it were familiar with its arguments through exposition by men like T. H. Huxley, that able protagonist of the new theory against highly placed but uninformed opposition. *The Descent of Man* attracted almost equal attention. The virulence of the controversy these books aroused gradually lessened as men either defiantly rejected the rational arguments of the evolutionists and clung to blind faiths, or else built the new ideas into their personal philosophy as best they could. Many people lapsed into agnosticism but the concepts of evolution did much to stimulate religious liberalism and bring the Church into line with modern science. The old ideas about man's nature, however, had been gravely shaken and in the minds of many he never recovered the exalted position that so long was his. Here was a fall of man as serious in its consequences as that other Fall which was a cornerstone of Christian doctrine. In losing confidence in the divinity of man, many

[1] *Ibid.*, Ch. III.

tended to lose confidence in Divinity itself. There can be no doubt that the theory of evolution was responsible for much of the reaction against religious orthodoxy which began even in the heyday of Victorianism and continues still.

Another challenge to the old ideas of man's nature gained momentum at about the century's turn. It came more gradually and was less dramatic but in the end became more devastating. This was the concept that whatever his origin, man is simply a physical and chemical mechanism and not different fundamentally from any other machine. The idea that he has a mind or soul distinct from his bodily activity it looks on as preposterous. The concept was not new, for materialism had long been supported by a respectable minority of philosophers, but through the years it had made little headway against the overwhelming forces of religious faith. What made it now so strong was the mass of new support for it which came from the young sciences of biochemistry, physiology and genetics. These are not concerned primarily with a study of man but with an application of the principles of the physical sciences to biological processes. They seek to analyze, in material terms, the activities of protoplasm and the systems that it builds. In this they have already made substantial progress and made it on the basis of one simple assumption—that there is nothing mystical or undiscoverable in man or any other living organism but that everything here follows the general laws exemplified in matter and energy elsewhere. These operate in the same way in living as in lifeless systems. This was a daring assumption to make, for biological orthodoxy had long maintained that in life there was something 'vital,' something specifically different, which set it apart from the rest of nature. To challenge this was obviously to strike deep.

This bold assumption has justified itself, however, by the great additions to our knowledge of living things, and man among them, which have been made through use of it. The long series of reactions heretofore thought limited to living creatures began to be successfully imitated in the test tube. The processes of metabolism were analyzed in chemical terms. There is no reason to believe, say most physiologists, that in the human body anything more is involved than the basic laws of chemistry with which we are familiar.

The principles of physics are also exemplified in living systems, and it was there that many of them first were studied. The basis of our knowledge of osmosis and of the permeability of membranes, for example, was laid by plant physiologists. Other physical processes are manifest in organisms as well as in lifeless systems. The science of biophysics, which is concerned with these problems, is certain to make great contributions to our knowledge of protoplasm and its conclusions emphasize those of biochemistry, that man's body is the seat of a wide range of physico-chemical processes.

Genetics is also involved in this interpretation of life in material terms, for it has discovered the physical basis of inheritance in the activity of definite minute bodies, the genes, of approximately molecular size. These are extraordinarily potent bits of matter and seem to control, step by step, the various processes that go on in any organism. This rigid genic determination, modified though it is by environmental factors, emphasizes again the conclusion that all living things, and man among them, are not independent systems but follow the immutable processes of physical law.

The bearing of these facts on the question of man's nature is obvious. If he is a physical system, however exquisitely complex, he must obey the laws of the physical universe and is as subject to their control as any machine. As to just how this control operates there are two somewhat different ideas. It may be thought of as coming chiefly from within, as in an automatic mechanism. This is the general view of physiology. Or the control may be thought of as primarily external, imposed on a rather neutral, plastic system by factors in its environment. This is a view commonly held by psychologists.

It is here that the science of behaviour carries still further the arguments drawn from the physical sciences and biology and applies them more closely to man himself. If man is a mechanism it ought to be possible to deal with him as we do with any other machine and to determine his behaviour as we wish. Many psychologists believe that this can be done and that the highest hope for man is in his being so wisely guided and controlled that he will become, so to speak, a custom-made saint who never does wrong nor ever wants to do so. Since his acts are determined not by himself but by forces outside himself, his leaders can take the control of human progress into their own hands. A man is an 'empty' organism, and everything he is and does is the result of environmental factors working upon him.

One of the leaders of this school of thought, Professor B. F. Skinner, put it thus: 'Just as biographers and critics look for external influences to account for the traits and achievements of the men they study, so science ultimately explains behaviour in terms of "causes" or conditions which lie beyond the individual himself. . . . Every discovery of an event which has a part in shaping a man's behaviour seems to leave so much the less to be credited to the man himself; and as such explanations become more and more comprehensive, the contribution which may be claimed by the individual himself appears to approach zero. Man's vaunted creative powers, his original accomplishments in art, science and morals, his capacity to choose and our right to hold him responsible for the consequences of his choice—none of these is conspicuous in this new self-portrait.'[1]

The techniques which science will use to control behaviour are those of 'supplying information, presenting opportunities for action, pointing out logical relationships, appealing to reason or "enlightened understanding",' and presumably more subtle forms of conditioning made possible by science. These are as much methods for the control of behaviour, says Professor Skinner, as the bully's threat of force. We have long recognized education and persuasion as means of getting people to do what we want them to do, though often with only partial success, but these new planners hope to go much further and make men virtuous and happy almost automatically.

This challenging proposal to make over human nature and solve the problems of a troubled world by a judicial application of the principles of psychology has a wide appeal. If men's minds could thus be moulded to ensure any desired pattern of behaviour, this would surely result in a radical change in the

1 B. F. Skinner, 'Freedom and the Control of Men,' *American Scholar*, 25:47—65, 1955-56.

lives of all of us, though whether or not it would usher in the millennium may perhaps be doubted. The idea presents some grave questions, however. If man is indeed the tractable and docile creature that the psychologist seems to think him, our ideas about him must be greatly changed. Where, we may ask, is now that freedom we thought he prized so highly? It would have to be surrendered; not, indeed, to physical determinism but to the edicts of a culture-planner or a brain-washer. And where goes moral responsibility, the concept of what 'ought' to be? What about values? What is the significance of man's passionate love of beauty, of the poet's rhapsody, the mystic's vision? Can proper conditioning confer the ability to create a work of art? And who would decide what sort of behaviour to inculcate? Such questions are not altogether frivolous. Whether or not man can be so readily controlled will depend on whether he is indeed an 'empty' organism, a blank sheet of paper, or has instead an inner quality that makes him resist environmental influences.

If life is mechanical, as both physiologist and psychologist believe, and thus is rigidly controlled, we may ask what place there is in an organism for mind, or any immaterial agency to guide its activities. This question of the relation between body and mind is one of the oldest with which philosophy has to deal and still remains unsettled. An automatic machine presumably does not have a mind, and mind in us may perhaps be only an illusion, a sort of by-product of physical acts; an epiphenomenon, produced by them as inevitably as the picture on a television screen is formed by its inner mechanism. The instinctive and commen-sense opinion that we have minds and that they control what we do is hard to reconcile with the mechanical view of man that science often takes. In trying to follow both these concepts today we are in danger of becoming hopelessly divided ourselves.

But if mind is hard to find a place for in the human organism, what of the soul? the spirit? human values generally? The challenge of physiology and psychology to the old ideas about man's nature is evidently far more serious than that of evolution. Man might come to think of himself as an ennobled animal, moving up the evolutionary stairway towards a divine nature, and thus regain some of his old assurance and sense of cosmic significance; but to think that he is simply 'an eddy in a stream of energy,' a mechanism buffeted by fate and circumstance and tossed at last into the rubbish heap—such a concept completely destroys his traditional ideas about himself.

Man's true nature is therefore the real problem—whether he is a child of God with an immortal soul and actually a part of the great spiritual power that rules all nature, or whether he is simply a clever brute, risen out of the primordial slime; a chemical mechanism that has evolved into a glorified calculating machine whose aspirations, seemingly so exalted, are nothing but motions among molecules, a puppet whose fate is no longer in his own hands. The serious division in the world today is not so much on social or economic or political issues as on this basic question of what man really is. Until we can agree on this we shall never be able to build a peaceful and satisfying social order for we shall have no common philosophy on which it can be founded. What man believes about *himself* is of the utmost moment, for it will determine the kind of world that he will make and even his own fate. This is the reason why those disciplines that deal with man are being explored so earnestly today. Anthropology and psychology have largely displaced theology as the centre of our interest. This is natural, for our concepts of God's nature are bound to be closely related to our concepts' of man's nature. Our understanding of this, in turn, will depend on what we know about living things in general. These deep matters come at last to focus in the question as to what life itself is. They are all problems of biology but of a biology which in breadth and depth covers the *whole* of life. We must understand *life* to understand man.

3

Life

IN THE quest to discover what man really is, the obvious course is to examine man himself. 'The proper study of mankind is man' has been the motto of most explorers of his nature. This is reflected in the vigorous development of the 'human' sciences today. Anthropology has come to be a major field of learning—the study not only of man's bodily evolution and diversity but of his cultures and their origin, of his development from barbarism to civilization. Sociology is the examination of his social systems, of how he gets along with other men and solves the problems caused by differences in ability, race, class and productivity. Psychology boldly undertakes to come to grips with what his mind is and how his conduct is determined; with behaviour, motivation and the self.

But the problem of man goes far deeper, I think, than man himself. The facts that told most heavily against the old ideas of what he is were biological facts from the fields of evolution and psysiology. Man is basically a biological problem. The questions we shall finally have to answer about him are biological questions. Those which anthropology and sociology ask will have to wait their complete solution until we understand something deeper than the nature of man alone—until we learn the nature of *life* itself. The final questions of philosophy are not about matter and energy and the cosmos, important as these are, but about the character and significance of life in a universe that is overwhelmingly lifeless. Lifelessness is the rule, life the puzzling exception.

Man is the noblest of the animals, we admit, but his high qualities are not his alone. They must have come from something further down the scale, for the great teaching of evolution is that the history of life is a continuous history. Until we find in the lower forms of life the roots of those great intangibles which reach in man their supreme expression—mind, will, spirit, soul—we shall not reach a final understanding of his nature. I do not mean that biology alone, without the aid of the human sciences, can learn all there is to know about mankind; but without biology a secure foundation for these other disciplines cannot be laid.

He who would explore the realm of man must for the moment, therefore, become a biologist and try to answer the question towards which all others in biology converge—the nature of life. Two contradictory answers have been given to it, corresponding to the two contradictory evaluations of men that have been stated here. Earlier biologists believed in the operation of a vital force or principle which governed the behaviour of living things and was related to the psychical element in man. They regarded what went on in organisms as different in character from the events in lifeless systems with their rigidly determined course. Hence came the distinction betwen 'organic' chemistry, which dealt with the products of living things, and 'inorganic' chemistry, for it was long believed that organic compounds could be formed only by living organisms.

The modern philosophy of mechanism discards this vitalism completely as mystical and unscientific. The uniformity of nature, its orderliness and dependability, are what make science possible, and science assumes predictability in the behaviour of *all* physical systems, and their conformity to law. This position must make one suspicious of all vitalistic ideas, for how could science survive half lawful and half lawless; bound by orderliness and predictability in the lifeless world but subject to a chaos of uncertainty in the living one?

There is more specific evidence for belief in the similarity of the living to the lifeless than this presumption alone. When the first organic substances were synthesized in the laboratory the ancient distinction between the two realms of chemistry quite broke down. Numberless characteristic products of living things can now be made outside their bodies. Quinine, for example, no longer depends for its origin on the cinchona tree or thiamine on green plants. These compounds and thousands like them are made now by synthetic chemists. Many more complex substances, like the proteins, are still beyond our competence. No one yet has repeated in a test tube the process by which carbon dioxide and water unite in green plants to form sugar, from which all food comes. There is no theoretical reason, however, why these processes should not be imitated, and he would be a rash prophet who would say that even the most complex compounds will never have their origin in the laboratory.

Still stronger support for the modern view is derived from detailed analyses of the activities of living things now made by biochemists and physiologists. The process of respiration, for example, by which energy from food is liberated to run the bodily machine, has been shown to be far more than the addition of oxygen to a molecule of glucose and its breakdown into six molecules of carbon dioxide and six of water. This is what happens in the combustion of sugar outside the organism, a process that takes place only at high temperatures. In 'the slow and smokeless burning of decay,' as Robert Frost calls it, or in our own bodies, this process goes on at a much lower temperature but with a fantastic complexity. The molecule of glucose step by step is broken down and after an elaborate cycle of changes nothing but carbon dioxide and water remains.

The important fact here is that biochemists have analyzed this process into terms of understandable chemistry. Nothing mysterious, nothing chemically unorthodox, here takes place. Not every step is yet understood, by any means, but the way is open for a complete exploration of this and other physiological activities in living things. Chemistry and physics have explained in material terms so much of what was once thought to be unexplainable that many of their practitioners now feel justified in assuming that everything about life can thus finally be accounted for. To the lawless frontier of metabolism law has finally come and this territory will never be the same again. It has lost the excitement of unpredictability but gained the great advantage of orderly and understandable behaviour.

Other processes, as well, have begun to be analyzed. Necessary for the successful conduct of many physiological activities are minute amounts of specific chemical substances, the vitamins. No more dramatic discoveries have been made in recent years than the proof that many serious diseases can be cured by the addition to the diet of a little thiamine or riboflavin or pantothenic acid. So impressed have people become with these wonder-working compounds that scores of millions of dollars' worth are purchased annually in the USA alone. In some cases biochemists have found just how the vitamin molecule fits into a vital reaction. Other substances, the hormones or chemical messengers, are also of great importance in physiological processes and bodily development, and their action is beginning to be understood.

Most chemical changes in a living organism are begun or controlled by enzymes or catalysts, minute in amount but very potent in their effects. Almost every reaction requires its specific enzyme, which may either break a substance down or synthesize it. Sir Charles Sherrington has graphically described how a living cell 'is like a magic hive, the walls of whose chambered spongework are shifting veils of ordered molecules, and rend and renew as operations rise and cease. A world of surfaces and streams. We seem to watch battalions of specific catalysts, like Maxwell's "demons," lined up, each waiting, stop-watch in hand, for its moment to play the part assigned to it, a step in one or the other great thousand-linked chain process. Yet each and every step is understandable chemistry. The cell has proved to be a perfect swarm of catalysts, or of trains of catalysts, each a link in a serial suite of chemical reaction.'[1]

The analysts of protoplasm are not all biochemists. Biophysics is destined to play an increasingly important part, for physical principles are exemplified in all living systems. Diffusion osmosis, absorption, membrane permeability, the transfer of electric charges, the operation of bioelectrical fields, the effects of various types of radiation—all these are manifest in organisms as well as in lifeless systems.

The upshot of all this is the conclusion of many scientists which is well expressed, again, by Sherrington. 'Natural science has studied life,' he says, 'to the extent of explaining away life as any radically separate category of phenomena. The cate-

[1] Sir Charles Sherrington, *Man on His Nature* (New York: The Macmillan Company, 1941), p. 78.

gories of living and lifeless as regards science disappear; there is no radical scientific difference between living and dead. . . . Living, so far as breathing, moving, assimilating, growing, reproducing, etc., amount to life, has by natural science been accounted for—some might say, "explained." There is nothing in them which does not fall within the province of science. They are chemistry and physics,'[1] He quotes Joseph Needham's remark that 'Biologists find their work is only possible if they define life as a dynamic equilibrium in a polyphasic system consisting of proteins, fats, carbohydrates, sterols, lipoids, cycloses and water.' A physical system made of certain specific substances is all that the physical sciences discover in life.

Here is the same antithesis, extending throughout biology, that is found in our conflicting ideas of man. If life is rigid, determined, mechanical, there is no room in it for those peculiarly human qualities that we so value. If, instead, we believe that mind and spirit and freedom are *real* and worthy of our allegiance, we seem to deny the basic philosophy of science and to rule out, as irrelevant to an understanding of man's deeper nature, the vast knowledge and resources that it has amassed and its magnificent philosophy of the orderliness of nature. This is a grievous dilemma and the seeming irreconcilability of its two horns is plaguing man most seriously today. There is a basic disagreement here that no amount of pious platitudes can heal.

We are sometimes inclined to accept this double standard for philosophy and see in the opposition between religion and science, intuition and reason, mind and body, vitalism and mechanism, evidence of a thoroughgoing doubleness in the universe—two different ways of looking at the same thing. Bohr's concept of complementarity, designed to account for the seemingly fundamental contradiction between classical and quantum mechanics, can be appealed to in justification of such a conclusion.

But man is too much of a monist at heart ever to be satisfied with a split through the middle of his philosophy. If he lives in a *uni*verse, it ought to be a unity, not a duality. There should be some middle ground between the obscurity and stagnation

[1] *Ibid.*, p. 291.

of vitalism and the rigidity of mechanism. Vitalism seems to disregard the uniformity of nature, but mechanism may have over-reached itself by taking too narrow a view of this great principle and sacrificed common sense to the requirements of an over-simplified theory. The only hope of reaching a reconciliation between these two positions is to find a concept of life that will satisfy both our deep convictions about it and the rational demands of science. The present philosophical impasse is so serious that the biologist should undertake to examine again, without prejudice or preconception, the nature of that remarkable phenomenon with which he deals and to find, if possible, the *via media* between these opposing views about it.

During the past half century there has been a movement in biology and philosophy in a direction which offers promise of finding this middle way and bringing matter, life and mind—concepts which have reality and value but are so difficult to reconcile—into a common system. These new ideas centre around the conception of *organization* as fundamental in biology, and of the organism as the great fact in the sciences of life. Aspects of this philosophy of *organicism* are evident in the work of Driesch, von Bertalanffy, Smuts, J. S. Haldane, Needham, Ritter, E. S. Russell, Ungerer, Schrödinger, Lloyd Morgan, Meyer Abich, Ralph Lillie, Whitehead and others. These men approach the problem from different points of view and none of them has been able to present the satisfying sort of explanation of organization that mechanism has achieved for metabolism, but the importance of the *idea* of organism has been growing in the minds of many thoughtful scientists and philosophers. Von Bertalanffy quotes Woodger's remark that the history of biology will some day include a chapter on 'The Struggle for the Concept of Organism in the Early Twentieth Century.' 'It will describe,' he says, 'how this idea was neglected under the influence of Cartesian philosophy; how a mechanistic metaphysics did not allow biology even to dream of organisms as anything other than swarms of tiny hard corpuscles; how the first appearance of the concept of organism at the turn of the century was frustrated by an inappropriate formulation, Driesch merely replacing the absurd notion of a machine without a mechanic by that of a metaphysical engineer; finally, how the concept of organism was first taken seriously not by biologists but by some philosophers and mathematical physicists.'[1]

Stiernotte remarks that 'this factor of organization which sustains the life of the whole organism is the central citadel of biological inquiry and must rank as a metaphysical principle.'[2] L. J. Henderson believes that 'organization has finally become a category which stands beside those of matter and energy.'

The study of organization has not been pressed far enough to answer the great questions we have been asking about man, but it is through further exploration of this region, I think, that the opposed conceptions of life—and of man—are most likely to become reconciled. Biology has gathered a vast amount of factual knowledge about the processes of organic development, but the point on this frontier where our best efforts to advance have persistently been repulsed is the question of how a formed and organized living system, the organism, is produced. Metabolism we are beginning to understand; growth, which is basically the production of new protoplasm out of food materials, seems to present no insuperable difficulties; but how a plant or animal develops from a single cell or a small mass of embryonic cells with all its parts so closely co-ordinated at every step that an organic system is produced and maintained is still beyond our comprehension. The problem is one with which we are all familiar as we watch plants and animals and children grow and observe the correlated behaviour of mature organisms. All this seems so natural that we often forget how astonishing it really is.

An egg does not merely expand, growing in all directions to form a spherical mass of cells. Almost from the start the embryo's growth is differential, more rapid in some directions than it is in others, so that even in its early stages a developing animal or plant acquires a precise *form*. In it there is a polar axis, different at the two ends—shoot and root, head and tail. Around this in a symmetry either radial or bilateral develop the various organs of the body—leaves, limbs and many others,

[1] Ludwig von Bertalanffy, *Problems of Life* (New York: John Wiley & Sons, Inc., 1952), p. 197.

[2] Alfred P. Stiernotte, *God and Space-time* (New York: Philosophical Library, 1954), p. 343.

each with a definite position and form of its own. Within, the cells become markedly unlike and the tissues and internal structures begin to differentiate in a definite pattern—pith and cortex, wood and phloem in plants, and muscles, nerves, secretory tissues and others in animals. During this time increase in size is taking place, and finally the young organism, moulded into its final form, is ready to start another reproductive cycle. All this may go on in open sea water, within an egg, inside the body of the mother or in the protected tissues of a bud.

A young embryo thus unfolding into an adult is one of the most fascinating sights in nature. Three notable facts distinguish it. First, it marches ahead in a regular and predictable fashion, one precise step after the other, to its final consummation, as though some unseen craftsman were fashioning it according to his design. Each part bears a close relationship, both in structure and activity, to every other part, from the beginning of development until the end, so that the whole is an organized system, well named an organism. Second, it possesses a very definite form and structure, so specific that almost every kind of animal or plant may be recognized and classified thereby. Morphology, the study of form, as Darwin said, is the soul of biology. The precise form an organism assumes is a visible expression of its organization, and the orderly developmental march is the means by which this comes to being. Third, if normal development is disturbed, there at once begins a series of processes which tend to restore it. Injuries are healed. Missing parts are regenerated. Altered patterns are reconstituted so that a whole and typical individual tends to be produced. The *self-regulating* capacity of organisms is often shown more dramatically in these modifications of their developmental progress than in normal development itself. In both, however, the essential fact is that growth is so controlled that an individual organism of a very definite character results. Organization is a regulatory process.

This fundamental quality of life is well stated by Russell, a zoologist: 'If in a living animal normal structural and functional relations, either external or internal, are disturbed, activities will usually be set in train that are directive towards restoring structural and functional norms, or establishing new norms which are adapted to the altered circumstances.'[1]

A few examples of this organized development, this self-regulated directiveness, will serve to make clearer what the process is like.

The development of an amphibian larva or 'tadpole' from an egg is a familiar part of a course in embryology. A frog's egg after being fertilized divides into two cells and these into four and so on until it has been cleft into a spherical mass of many cells. These pull apart to form a hollow sphere. One side of this becomes pushed in, and the hollow thus created is the primitive digestive cavity. An opening, breaking through at the other end, will make the mouth. Above the back of the tiny embryo the surface puckers lengthwise into two folds which arch toward each other to meet and enclose what will be the spinal cord. Gills grow out on either side of the mouth. Two little buds at the front and two behind develop into the four limbs, and the rear of the embryo draws out into a tail. While these external changes are taking place, a precise pattern of inner organs and their tissues begins to appear, and finally the larva breaks out of its gelatinous matrix and swims off. In time the gills and tail disappear and the tadpole grows into a mature frog. This is the general plan followed in the development of all vertebrate animals. The human embryo itself, though producing a much more complex adult, passes through a very similar series of stages.

The developmental history of the frog, familiar in its outer features to almost every schoolboy, is marvellous enough, but it seems even more so when one interferes with it and observes the resourcefulness with which it nullifies his efforts. When the egg divides into two cells by a partition wall, it is possible, with care, to separate these cells from each other. Each now proceeds to develop independently, but instead of forming half a larva, as it would have done if left in contact with its sister cell, it forms a *whole* one. Its entire growth programme has been rearranged. How can it be, we ask, that the fate of each cell is now so very different? The ingenious experimenter can go still

[1] E. S. Russell, *The Directiveness of Organic Activities* (Cambridge: Cambridge University Press, 1945), p. 44.

further. By the proper technique two separate eggs may be fused to form one cell and will then go on developing. What grows from this double egg is not a double larva but a single one, entirely normal except for its larger size.

There are many other cases where a single cell, if isolated from the rest, will form a perfect whole and not a fraction of one. The means for making a whole are obviously present in every cell. In the developing embryo, what a particular cell will produce depends on its particular place in this whole. As Driesch put it, 'The fate of a cell is a function of its position.'

The growing tip of a plant shoot is a perpetually embryonic region, the site of a continuing series of embryonic histories as organ after organ is formed and develops to maturity. The tip itself is a minute dome-like group of cells about as large as the point of a blunt pin. Here the cells are increasing in number by division, and growth and development are beginning. Leaves, floral parts, buds and other organs arise as small projections growing out below this rounded tip, and it is in this region that the general plan and symmetry of the shoot are established.

In a maple bud, for example, just below the terminal dome, five tiny bulges appear on each side of the young axis, opposite each other. These grow in length to form the finger-like ribs of two minute leaves. Other leaf rudiments appear and soon the entire next year's shoot in miniature is formed. In this state the bud passes the winter. In the spring growth once more begins and each tiny five-fingered leaf primordium expands, both by increase in the number of its cells and in their size, into a full-grown maple leaf. Within the paper-thin blade veins appear in a precise network, grading down from the five large ones to their final minute branches. The green cellular tissues of the blade form two layers, one next to the upper surface, dark green and closely packed, where sunlight and chlorophyll co-operate to make sugar; and the other below, with ample air spaces between its cells, where gas exchanges between leaf and atmosphere take place. At the base the leaf stalk bgins to grow and carries the blade outward at its tip. This developmental story is repeated thousands of times on a single tree each year and follows the same precise and orderly course. Something in the minute primordium in the bud says 'maple leaf,' and not the leaf of any maple but of sugar maple or silver maple or Norway maple.

The development of a flower is still more complex. Around the tip of the minute dome that is to give rise to a flower bud appear circles of knob-like outgrowth such as those which make a leaf. The outer ones develop into sepals, the next into petals, the next into stamens and in the centre appear the ovary and its parts. In many species, as the bud enlarges, these various floral structures are packed and folded and twisted tightly together, much like the tissue of a parachute but in a far more complex fashion. In the bud of an iris, just before it opens, one finds the three petals packed closely in a counter-clockwise spiral, the three stamens and three stigma lobes tucked inside them and the whole firmly enclosed by the sepals. When the bud opens all these parts unwrap themselves and open out into the iris blossom. The opening of an umbrella is simplicity compared to this. How these packed and folded parts grow from their minute beginnings so precisely that in the opening of the leaf and flower nothing is missing or is wrongly placed is one of the wonders of development.

In plants, even more than in animals, pieces of tissue or even single cells, if isolated from the rest of the individual, show an ability to restore the whole again. The power of cuttings to strike root has been used since the dawn of horticulture to produce new plants abundantly from old ones. In several species, notably the 'live-forever,' tiny plantlets grow out from the notches in the leaf margin, and finally drop off to make new individuals. Some species like the African violet are regularly propagated from leaf-borne plantlets but these arise from the surface of a leaf which has been cut off and placed on damp sand. Microscopic examination proves that certain cells in the outermost layer of the blade begin to grow, and sometimes it can be shown that a new plant comes from a single one of these cells. In many other cases cells from various regions of the plant, or groups of similar cells, have been shown to have the power, under favourable conditions, of producing whole plants by such a process of self-regulation. This does not always happen, for development has sometimes gone so far that a tissue cannot become embryonic again; but in the more primitive animals and plants and in the earlier stages of development this power of regeneration is very common. Often, as Driesch and others have pointed out, the restoration of a lost part or a disorganized pattern may be attained by different routes, as though the final goal and not the means of reaching it were the important thing.

Many other examples might be given of this tendency for development to march on to a definite goal. Some of these the present author has described elsewhere.[1] Thousands of tiny separate amoebae of a slime mould will come together in an aggregating mass, from which a precisely formed fruiting body is produced. If a living sponge is separated into its constituent cells by being squeezed through muslin, these will slowly come together and restore the sponge again. The tissues of a caterpillar, almost completely broken down inside the shell of the dormant pupa, are entirely reorganized to form the body of the moth or butterfly that emerges from it. The branches of a young tree, tied into abnormal positions, will tend in their growth to restore the pattern of the whole again.

Worthy of special mention, perhaps, are those cases where parts of two plants become fused, as in the process of budding or grafting so commonly practised in horticulture. Here the growth of each complements the other and they form a normal whole, though its parts are genetically unlike. More remarkable are those *chimeras* where there is a still more intimate mixture of the tissues of the two components. If a tomato scion is grafted on a nightshade stock, for example, shoots from near the graft union occasionally arise in which the outermost one or two cell layers come from the nightshade plant and all the rest of the shoot from tomato tissue, or the reverse. These two species differ greatly in the shape of their leaves, the size and colour of their fruits and other respects. The cells of the two, however, thus so intimately associated, do not produce a series of bizarre abnormalities but make the best of it, so to speak, and join forces to form a normal plant, though this is unlike either nightshade or tomato. The tendency to organize a whole is so strong that it succeeds even under these unusual conditions.

In the activities of a plant or animal a similar self-regulation

[1] *The Biology of the Spirit* (New York: The Viking Press, 1955), Ch. II.

is evident in the maintenance of a particular 'steady state' of body temperature, blood sugar or many other things. Such *homeostasis* is the final problem of physiology. 'The living structure is evidently organized: that is to say every part of it bears a definite relation to every other part. As, however, the structure is the outcome of metabolic activity, it follows that the metabolic activity of the living body is also organized, every aspect of it bearing a definite relation to every other aspect. That this is actually so has become more and more clear with the advance of physiology, particularly in recent times.'[1]

These various examples point to a fact of much importance in the problem being considered here, for this integrating, co-ordinating ability by which the activities of protoplasm are so regulated that it builds an organism seems to be the essential quality of life. Each part of the developing body tends to form what is appropriate for its position in the organized system. Something representative of the whole organism, so to speak, is present in every cell and guides that cell to make its necessary contribution to the whole. In confirmation of this conclusion is the fact that every cell of the body usually possesses the entire complement of chromosomes and genes which were present in the egg from which the body grew.

This quality of directive self-regulation, whatever its final relation to chemical and physical processes may prove to be, is a uniquely biological phenomenon, and an understanding of it, I believe, will provide a clue to the character of life itself. To explain it presents a problem of extraordinary difficulty and one must admit that little progress here has yet been made, a fact that often embarrasses botanists and zoologists, for it continues to lurk like a skeleton in the biological closet.

This characteristic behaviour of living stuff was particularly emphasized by Driesch more than half a century ago, who mde it the basis for his concept of vitalism. He explained it by hypothetically installing a little entelechy in the growing organism, which was able in some way to suspend physical law just enough to permit self-regulation. The opposition of mechanists to such an idea was based on the immutability of

[1] J. S. Haldane, *Mechanism, Life and Personality* (New York: E. P. Dutton & Co., Inc., 1914), p. 78.

natural laws, which vitalism seemed to deny. What the mechanists tended to forget, however, and the vitalists not to emphasize enough, was the unexplained fact of the *organism*. This is a fact, however difficult it may be to account for, and it is this fact that the philosophy of organicism has done well to place in a conspicuous position. Neither vitalism nor mechanism by itself gives a complete interpretation of life. Organization—self-regulation—must be accepted as a fundamental scientific reality. Whatever its explanation may be, it has important implications for the problem to which we have addressed ourselves. It opens up a road which need not be the dreary one of rigid determinism or the unscientific one of mysticism but which has the possibility of combining the qualities of each. Search for an understanding of man's nature must begin with an exploration of the biological frontier along this middle road that leads to the most distinctive quality of life. Here we shall discover a foundation, I believe, on which to build a philosophical structure worthy of our theme.

4

Life and Mind

WHATEVER biological organization may turn out to be in terms of matter and energy, genes and protoplasm, the *fact* of it is indisputable. The development and physiological activities of animals and plants are under regulatory control, explain this in whatever way we can. The implications of this fact are of the greatest significance in a study of man's nature for it is here, I believe, that the psychical touches the physical, the mental comes from the developmental, and mind emerges from matter. Let us examine the arguments for this conclusion.

When dead matter, random and fortuitous in its distribution, enters a living organism, it comes under the control of a regulatory system which moulds this hitherto disorganized material into a complex organic pattern of a very precise sort. Here the fate of every particle is the result of its position in the whole. Something presides over the growth of this living architecture and maintains a balance among its activities. The moment life departs from this body a radical change ensues. Chemical chaos now begins again; chemistry of the sort we understand, to be sure, but where each change is no longer part of an organized pattern, or in a precise relationship to all the others that are taking place

This organic pattern has a hereditary base, and its roots are in a protoplasm which is specifically different in the various kinds of animals and plants. It is built up by the interaction of thousands of separate genes, each of which, as modern genetic research has shown, controls or affects some process. How this throng of different factors co-operate so precisely that the body of an animal or plant is formed is a basic problem in biology. What is produced is necessarily affected by outside conditions, too, for what a gene controls is a specific mode of reacting to a specific environment.

This organized system, maintained by the regulatory control of its activities, implies the presence within it of *something to which these activities tend to conform*, a norm, a standard, a goal or end, what the philosopher would call a *telos*, inherent in the whole living mass. Regulation to this norm bears such a close semblance to *purposive* action as to suggest the universal presence of a kind of teleology in organisms. Most biologists have hesitated to recognize the implications of this idea. Purpose, they feel, is something that has no place in science, for it not only suggests an effect of the future on the present, and thus violates the ordinary concepts of causation, but it also assumes that a psychical factor can control material processes. Biology has so striven to gain status as a true science that to introduce into it this idea of purpose seems to many to push it back into mysticism again. The sort of purpose here to be discussed, however, is activity which tends toward the realization of a developmental pattern or goal present in every living thing. Such teleology, far from being unscientific, is implicit in the very nature of organism. The biologist need not shudder at these words, for the purposiveness of this sort is not only unobjectionable in his science but lies at the very heart of life itself. From this fact conclusions of the utmost importance may be drawn, for mind itself, I think, here finds its relationship to matter. If true, such a concept is significant not only for biology but for psychology and philosophy, as well.

The most cogent evidence for it is that self-regulation and goal-seeking are to be found not only in bodily growth and development but also in those other activities of protoplasm which govern its *behaviour*. Here the goals are the attainment or maintenance not of particular structures but of particular states. At the lowest level behaviour is purely physiological and involuntary, as in digestion and most glandular functioning. Activities like breathing are partially under conscious control and directed by the brain. Still higher are those acts by which an organism so regulates itself that it stays in a favourable environment, obtains its food and reproduces itself. These directives are concerned with the control of what the organism does. Just as there is a developmental goal immanent in the cells of the growing body, so we may infer that there is a goal of behaviour immanent in the cells of the nervous system. Such a goal that regulates behaviour is commonly called an instinct. Psychologists have debated whether there really are such things as instincts, how many there are, what their physical basis is, how far they are inherited and what their relation to intelligence may be. It seems clear, however, that most animal behaviour below the highest level, whether it is called instinct or not, is regulatory and tends to conform to certain goals. This was emphasized by William McDougall when he said that 'psychic activity is always and everywhere teleological, a striving toward some end.'

This is an inference of considerable moment. Here, co-extensive with life itself, is the simplest sort of psychical process, the dim beginnings of mind. Instincts—purposeful behaviour—are psychical traits. They are manifestations of life which link development and bodily traits with all the complexities and possibilities of mind. In behaviour, protoplasmic purpose grows to instinct, and with dawning consciousness this leads to thought and the higher elements of mind. Exactly what this process is that guides all vital activities to ends cannot be determined until we find *what the protoplasmic basis of goal-seeking turns out to be*. The problem is at bottom a biological one.

A developmental goal leads to the formation of a bodily structure in which, so to speak, it becomes immobilized in a static and material form. A goal to which behaviour conforms, however, since it is concerned with the activities of an organism, is much more fluid and may change frequently as the environment is altered. This, together with the fact that it usually involves motion, often hides the resemblance between behavioural and developmental goal-seeking.

Instincts are amazing things, as any naturalist well knows. I once watched a hummingbird build her tiny nest. She went about it in as skillful a way as any human builder. A few pieces of lichen were fastened to a branch by her saliva. Others were added and tied down by very delicate fibres—fern strands or cobwebs. More lichen was added and then in the centre various kinds of downy plant stuff. The little builder brought in bit after bit of new material, tucking it in at just the right spot, so it seemed. She pulled the fibrous strands from the outside up over the edge to anchor the whole, and in the end it was a wonderfully soft little cradle, hardly bigger than a walnut shell, for her two eggs. Her behaviour throughout was remarkably like that of a human craftsman, sizing up a particular problem, trying this means and that, and finally reaching a satisfactory solution. And yet this tiny bird had probably never seen a nest built, had nothing to imitate, had never learned the mechanics of nest building. This was perhaps the first nest she had ever made, and it was no different from those she might make in the years to come. There was something in her brain which represented the nest and all the steps in its construction and which, by a sort of trial and error, she sought to realize. It was something, I believe, not radically different in character from what was present in the egg out of which she grew and which represented the body of a hummingbird and the steps in its development.

Where, asks Bergson, can one draw the line between the constructive processes of the bird's embryonic development and the constructive process by which it builds its nest? Both are activities of protoplasm; one simply continues what the other has begun. In each there is immanent in the living stuff of the organism a goal or pattern towards which it moves; in one case expressing itself in bodily structure and in the other in bodily behaviour. Both are expressions of that goal-seeking purposiveness which is life's distinctive mark.

Such is the argument that behaviour, the simplest sort of psychical activity, is protoplasmic purpose grown to instinct, and finally to other aspects of mind. Conscious mind is the subjective side of this puposiveness, the inner experience of protoplasmic regulation at its highest level. Mind itself, at least in essence, seems to be coextensive with all life and grows out of that self-regulation and goal-seeking which is life's distinctive quality. Life and mind essentially are one.

Purposes first expressed themselves directly in action. As mental life grew more complex, action often was delayed and purpose was softened to a desire which might not be fulfilled. Finally action became largely mental and abstract thought was possible. Thought itself is basically a teleological process. Thus one may trace the origin of ideas from regulatory purposes and these, in turn, from a basic quality in living stuff that governs its activities. 'Biology is obliged to assert,' says Henderson, 'that ideas, whatever the philosopher may think of them, *at least* have a function, and that function, physiologically considered, can only be to regulate action.'[1]

Mind, in lower organisms, is a relatively simple affair. In man, however, greatly enriched by memory and imagination, it has become increasingly complex. Mind, of course, is far more than the seeking of a simple goal, for goals change as environment is altered. Even developmental goals may be very different under different conditions. A mermaid-weed has finely dissected leaves if growing under water but merely toothed ones in air. The larva of a bee fed on queen's jelly will grow to be a queen, but ordinary food produces simply a worker. Children given a complete diet will grow to be larger adults than those whose food lacks some of the requirements for full growth. Under each environment the goal is different, but each is normal under the conditions.

Behaviour in the same way is altered by environment. Increasing length of the spring days, for example, through its effect on the sex glands of some birds, so stimulates their instinct to migrate that this exceeds all others and they take off for the north. The environment of most animals is constantly changing. Their purposes—their primitive 'ideas'—are thus continually being altered. In a relatively sluggish species the behavioural goal is constant or alters only slowly. In an active one whose environment is continually changing, minor goals change, as well, though all activity tends towards the basic goal of normal life for the individual. In man with his complex inner changes and the wide range of external conditions to which he can respond, the stream of mental life becomes infinitely more varied than in any animal. Mental life is therefore to be regarded, I think, not simply as a 'stream of consciousness' but as a succession of goals that are being sought, of ideas that basically are purposes. In higher animals

[1] L. J. Henderson, *The Order of Nature* (Cambridge: Harvard University Press, 1917), p. 99.

these are set up in the nervous system, and an inner experience of them grows to conscious purpose, but their germ is present in all regulatory organic activity, even that of plants and the simplest animals. Mind need not imply consciousness.

Thus to draw purpose from protoplasm will appear to many an idea of little promise. To derive mind from the simple processes of growth and development seems to leap across a gap so wide that the attempt can hardly be expected to succeed. Nevertheless, the phenomena which seem so far apart in their full expression are actually close together at their protoplasmic beginnings and have many points of similarity. If one is to avoid the dualism of mind and body and find a common denominator for them both, the suggestion here presented seems to offer a hopeful possibility. If sound, it has far-reaching implications for an understanding of man's nature.

The idea of the purposiveness of vital processes is by no means a new one. It was held by many early biologists, but the reaction against anthropomorphism in science became so strong that discussion of 'purpose' in organisms often became taboo. The concept that the physical and the psychical may be united through their common origin in biological purposiveness is not widely held today, but it has a distinguished minority of supporters among biologists, though not all of them would state it in the same terms. Professor E. S. Russell comments on it thus: 'We recognize the fact that organic activities as manifested by organized unities such as cells and organisms, show characteristics, especially in their directiveness, persistency and adaptability which are shown also in instinctive and intelligent behaviour of ourselves and other animals. . . . Purposive activity, as seen in its highly developed form in the intelligent behaviour of man, is a specialized and elaborated kind of directive activity, concerned mainly with the mastery of his material environment. . . . Morphogenetic activities and instinctive activities are linked, complementary and functionally equivalent, and resemble one another in their essential characteristics.'[1]

Ralph Lillie goes deeply into the philosophical side of the

[1] E. S. Russell, *The Directiveness of Organic Activities* (Cambridge: Cambridge University Press, 1945), pp. 179, 181.

problem but supports the same conclusion. Says he: 'Conscious purpose, as it exists in ourselves, is to be regarded as a highly evolved derivative of a more widely diffused natural condition or property, which we may call 'directiveness.' . . . In the characteristic unification of the organism an integrative principle or property is acting which is similar in its essential nature to that of which we are conscious in mental life. . . . Conscious purpose is to be regarded as only one form of biological integration; the integration shown in embryonic development is apparently unconscious, and the same appears to be true of most physiological relations. Such biological facts point to the existence of a more general integrative property or activity of a fundamental kind which is universally present in living organisms, from amoeba to man.'[1]

J. C. Smuts discusses the matter as an aspect of his philosophy of holism. 'Mind,' says he, 'is a continuation, on a much higher plane, of the system of organic regulation and co-ordination which characterises Holism in organisms. Mind is thus the direct descendant of organic regulation and carries forward the same task.'[2]

Spemann sees the resemblance between physical and psychical events. In the last paragraph of his book on embryonic development, he says: 'Again and again terms have been used [in this book] which point not to physical but to psychical analogies. This was meant to be more than a poetical metaphor. It was meant to express my conviction . . . that these processes of development, like all vital processes, are comparable, in the way they are connected, to nothing we know in such a degree as to those vital processes of which we have the most intimate knowledge, viz., the psychical ones.'[3]

Bergson sees the relation of instinct to organization. 'Most instincts,' he says, 'are only the continuance, or rather the consummation of the work of organization itself. Where does the activity of instinct begin? . . . We cannot say, as has often been

[1] Ralph S. Lillie, *General Biology and Philosophy of Organism* (Chicago: University of Chicago Press, 1945), pp. 196, 201.

[2] J. C. Smuts, *Holism and Evolution* (New York: The Macmillan Company, 1926), p. 224.

[3] H. Spemann, *Embryonic Development and Induction* (New Haven: Yale University Press, 1938), p. 371.

shown, where organization ends and where instinct begins. When the little chick is breaking its shell with a peck of its beak, it is acting by instinct, and yet it does but carry on the movement which has borne it through embryonic life.'[1]

L. J. Henderson emphasizes the efficiency of instinct: 'It may, perhaps, be said that in performing such functions as adjusting the regulatory processes, cells seem to act as if they were controlled by something which remotely resembles intelligence, but which is really far superior in efficiency in that it operates necessarily, according to the needs of the moment, without the guide of previous experience, and without those errors of judgment which are all too common in voluntary action.'[2]

Occasionally a psychologist, though hardly an orthodox one, has come to the same conclusion that has been defended here. C. S. Myers, for example, denies that there is any *relation* between mind and body—any interaction between them, any 'psychophysical parallelism'—because mental activity and the activity of the living body are essentially the same. 'Mind and life,' he says, 'are identical properties of what we term living matter—a peculiar form of activity, differing from that of lifeless matter, in the inherent purposive, self-directive, and finally purpose-felt struggle for existence.'[3]

A position similar in many respects to the one presented here is taken by E. W. F. Tomlin[4] in a stimulating recent book. He approaches the problems of life and mind from the viewpoint of a metaphysician, and stresses the significance of biological values in these questions. What he calls values seem to be essentially what have here been termed goals.

There are others who might be cited. To develop at all adequately the idea that man's mental life comes from protoplasmic goal-seeking and to answer the objections that may be

[1] Henri Bergson, *Creative Evolution*, Tr. by Arthur Mitchell (New York Henry Holt and Company, 1911), pp. 139, 165.

[2] L. J. Henderson, *op. cit.*, p. 87.

[3] Charles S. Myers, *The Absurdity of Any Mind-Body Relation*, Hobhouse Memorial Lecture, 1932 (Oxford University Press), p. 9.

[4] E. W. F. Tomlin, *Learning and Knowing* (London: Faber and Faber, 1955).

raised against it is impossible in this brief space. The author has elsewhere attempted to do this somewhat more fully.[1]

This conception is far from being merely an interesting speculation. It carries implications of such importance for psychology, philosophy and even religion that they deserve to be explored more fully. If true, it offers a number of advantages. Not only does it give a reasonable foundation on which body and mind can be brought together, but it provides a basis for man's inherent and indomitable purposiveness, his motivation *toward* something, his concern with the future, his *intentions*. It also offers suggestions for other psychical aspects of his life. No seeker for an understanding of man's nature should fail to explore this promising frontier.

[1] *The Biology of the Spirit* (New York: The Viking Press, 1955). See also Chapter VII of the present book.

5

The Philosophy of Organism

THAT the mental grows out of the developmental and that the two are aspects of the same protoplasmic process is obviously an idea of great importance for our objective of reaching an understanding of what man is. It rests on the basic fact of biological organization, the existence of a purposive factor in both development and behaviour that makes protoplasmic activity conform to norms or goals. The difficult task is now before us of interpreting as best we can, in scientific and philosophical terms, this distinctive character of living stuff and thus finally, perhaps, of man himself. Here is a battle ground where mechanism, vitalism and organicism have locked horns for many years. Before it is possible to develop further this relation of the mental to the physical, and its implications for various problems of man's nature, we should therefore discuss briefly some of the ideas that are implicit in the fact of the self-regulating, organizing quality of living things.

First, the concept of purpose itself. As the word will here be used it has quite a different meaning from the sort of 'purpose' sometimes cited in the careless teaching of biology—the 'purpose' of a moth's long tongue to reach the bottom of deep flowers or the 'purpose' of the leaf to manufacture sugar. There is no evidence that traits of this sort are the result of any attempt by the organism to meet its needs, to adapt itself to its environment; any striving to meet the requirements of its surroundings such as Lamarck postulated as part of his theory. The host of adaptations which so long have impressed naturalists, such as the modifications of floral structures which favour pollination by insects, the protective colouring of birds and mammals, and the ability to form antibodies against bacterial toxins, seem to have arisen through the selective survival of favourable variations during evolution. The misuse of the word 'purpose' to explain such traits has brought the word into such disrepute that a biologist hardly dares utter it today. As here employed, the word means something very different—an organized pattern in protoplasm to which development and behaviour tend to conform. The very fact of regulation implies something to regulate *to*, a norm or goal set up within the organism. *Whether the attainment of this goal is advantageous or not is quite irrelevant.* Most new mutations that arise in animals and plants are harmful, as any random change in a precise mechanism might be expected to be. The short-legged Ancon sheep, for example, was useful to man in a country of stone walls but would have been seriously handicapped in nature. Doubleness in flowers, very ornamental but providing poorly developed stamens and pistils, is another case, and these could be multiplied indefinitely. All such mutants would be eliminated by the strict selective action of nature. They are, nevertheless, organized, self-regulating systems, though the goals they achieve are harmful to themselves. Most goal patterns now present in nature are advantageous to the organism, of course, since only these have been able to survive.

There is a fertile source of misunderstanding here, for one easily slips into the idea that the organism has a *natural ability* to *react favourably*. Von Bertalanffy, for example, distinguishes sharply between organization, the tendency to form or restore wholes, on the one hand, and purposiveness, on the other. The latter, he says, implies will or aiming at a goal and this leads to the false idea that an animal is 'trying' to do what a human being would do—act in a favourable way, its goal success and survival. This is *not* what happens in an organism at all, though selection has necessarily resulted in patterns of behaviour that are favourable. The most obvious meaning of purpose is activity toward an end, and this is what organic activity is. The term purpose refers to any goal-directed act, 'good' or 'bad.' An organism, like a person, does what it 'wants' to do, in the widest sense of that term. I cannot see why organic activities of this sort should not be looked upon as similar in essence to human purposiveness, though apparently unconscious. This is the present argument. If one clearly understands that such organic purposiveness does not involve intelligent choice of what is known to be favourable (which would be false anthropomorphism, since the lower organisms are not rational), the use of the term 'purpose' here is entirely defensible. It calls attention to a basic similarity between the behaviour of man and simpler types of living things, a *true* anthropomorphism.

This sort of organic purposiveness or teleology is implicit in the idea of a self-regulating organism. It need not be involved at all with what Aristotle called 'final' causes, a mysterious influence on future events of a conscious purpose or desired end in the mind. Such seems to imply an effect of the future on the present and is opposed to scientific ideas of causation. 'Purpose' of the sort considered here may be present in any automatic mechanism, living or lifeless, that regulates its activity to a norm or goal. Claude Shannon's electronically guided mouse which has been 'taught' to follow a precise path through a maze may be thought of as having within it a purpose of a sort. An automatic steering device on an aeroplane can be set for a given course and will bring the plane back to this course whenever it begins to deviate. A simple thermostat keeps a room at 70°. In none of these cases do we say that the future affects the present or that any 'final' cause is operating. There is simply an automatic, self-regulating mechanism set for a given goal. This mechanical behaviour is not generally called purposive because mind presumably is lacking in it, but one can imagine the regulatory activity of a living organism to be controlled in some such way as in these mechanisms. Whether such activity can be explained on physical principles with which we are familiar or involves others that are still unknown is a problem for physics and philosophy. Certainly no protoplasmic system we have yet discovered can account for this regulatory behaviour. My point is that there is nothing inherently contradictory in a machine that is teleological. Indeed, any mechanism may be said to have a purpose in this sense. The 'purpose' of a machine, however, was first in the mind of its designer and did not arise by itself, as it does in an organism. As to just what the nature of purpose in an organism is—whether it has a purely mechanical origin or involves something deeper—therefore depends on what the protoplasmic basis of biological self-regulation finally turns out to be.

Nor is the sort of teleology here discussed involved in the question of ultimate purposiveness in nature, of orthogenesis or directive evolution. Whether such occurs or not is still a question, though most evolutionists today see little evidence for it and prefer to regard the history of organisms as the result of chance variations. This is a difficult problem and certainly cannot be solved until our knowledge of evolution is much more complete. Universal teleology is a still deeper question, about which L. J. Henderson's speculations are interesting. He calls attention to the remarkable qualities of the elements carbon, hydrogen and oxygen, qualities without which life could not exist, and suggests that there may be a preparatory relation, so to speak, between the nature of the physical universe and the nature of the life that has developed in it. 'The appearance of harmonious unities in nature,' says Henderson, 'which no man can escape, depends upon a genuine harmonious unity that is proved to exist among certain of the abstract changeless characteristics of the universe.'[1] None of these more metaphysical aspects of teleology is concerned in the sort of organic purposiveness we have been discussing.

What *is* involved is the problem of the nature of biological organization. The ability of mechanistic philosophy to explain developmental regulation was challenged by Hans Driesch more than half a century ago. He showed that in early states of the embryology of certain animals a single cell or group of cells, if isolated from the rest, is able to produce a whole individual. Such a *harmonious equipotential system*, as he termed it, has now been found to occur in other animals and in many plants. This entirely discredited Weismann's earlier hypothesis that all the potentialities of the individual, present in the egg, are parcelled out to the various cells as development takes place, for *every* cell seems to have *all* the possibilities that were in the egg itself; and among plants many cases have been found where whole new individuals can actually be produced from single cells of the body. The presence of the entire organism in

[1] L. J. Henderson, *The Order of Nature* (Cambridge: Harvard University Press, 1917), p. 206.

every cell, so to speak, is the basis for the idea that in the fertilized egg and the cells that develop from it there is immanent something that can be regarded as the simplest manifestation of a psychical trait, a purpose towards an end. Here is true teleology but without the metaphysical concept of a 'final' cause.

That something like a goal or purpose is involved in the self-regulatory process of development is supported by the remarkable fact that the restitution of lost parts (and regulation in general) may proceed *by very different paths* but still achieve the same result. A classical case of this is the restoration of the crystalline lens of the eye in the marine snail *Triton*. Here in normal embryonic development the lens is formed from cells of the surface layer which make the skin, but the regenerated lens is produced by the optic cup arising in the brain, a very different organ and tissue. Many such cases have been observed in a wide range of organisms.

In instinct the same fact is evident. Says Agar: 'The chief objective indication of purposiveness in the behaviour of living things is the familiar fact that the sequence of acts by which the goal is attained is not always the same. On different occasions the organism reaches the same end by different routes. . . . Even in the most rigidly instinctive behaviour the animal must always fit the details of its action to the special situation. The completed nest, the spider's web, the act of mating, is attained by a train of acts different in detail on every occasion.'[1] All these facts suggest that it is the *goal*, rather than a fixed series of successive steps, that dominates development and behaviour.

Experimental morphologists have long been trying to find the basis for such regulatory activity. The development of complex electronic calculators has given hope that something similar to them might be present in protoplasm. Norbert Wiener calls attention to the resemblances between the two, and to the fact that the so-called 'feed-back' mechanism provides a means for correlating the activities of a system. Whether this is more than an analogy, however, is debatable. Servo-mechanisms and systems of coding which apply the

[1] W. E. Agar, *A Contribution to the Theory of the Living Organism* (Melbourne: Melbourne University Press, 2nd ed., 1951), p. 21.

results of so-called 'information theory' may well throw some light on the regulatory processes of protoplasm.

The principle of Le Chatelier, that a dynamic system, when modified by stress, tends to restore itself, has also been appealed to as a possible basis from which organic self-regulation may have been derived and is worth considering as a link between organization in the living and the lifeless. Even Pauli's discovery that the orbit of a given electron is precisely related to those of all other electrons in an atom has been regarded by some as one of the simplest instances of what might be called organization. The fact that the organism is an 'open system,' which persistently maintains itself but through which matter is continually pouring, has led to an intensive study of such systems and an attempt to formulate a general theory for them by von Bertalanffy and others. Bio-electrical fields which show marked constancy in their patterns of potential differences also seem to be concerned in some way with protoplasmic organization.

None of these ideas has gone any considerable distance towards explaining the complex phenomena of organic self-regulation, but they have suggested clues which may in time lead to a clearer understanding of the relation between the physical basis of life in protoplasm and its characteristic self-regulatory and goal-seeking behaviour. A few philosophers would have us take the fact of organization—and of life—as something *given*, something to be accepted like energy as a basic and unexplainable fact of nature. We may in time have to do this; but such a defeatist attitude tends to discourage scientific attempts to understand what organization really is. It must be admitted that the intensive research of half a century has yet done little more than formulate the issue in clearer terms. But I believe the problem is by no means hopeless. To solve it is the ultimate task of biology, and every effort will certainly be made to do so. Research from many directions may be expected to throw more light on it, though perhaps in ways that are now quite unexpected. There is something more in protoplasm than yet meets the eye, and our respect for its complexity has greatly increased in the past few decades.

The difficulties of picturing the origin of a self-regulating living mechanism are great. The building of a lifeless machine one can observe and understand. If he watches an automobile slowly taking shape on an assembly line he sees part after part added—chassis, wheels, motor, body, each with its specific details of structure. Finally it is completed, filled with gas and ready for the starter's touch. In its construction no physical laws are violated. Indeed, they are utilized to make the construction possible. A complete inventory could be made of the physical and chemical changes involved and of the energy expended by the machines and by the bodies of the workmen, and this could be described in purely physical terms. An observer who should make such an inventory without knowing anything about the manufacture of automobiles would doubtless be impressed with the complexity and orderliness of the process. It would be entirely incomprehensible to him, however, until he realized that it was the working out of a plan in the mind of the designer. An automobile serves a purpose but this is the purpose of its builder, and such self-regulation as it possesses is imparted to it from without. An organism, on the contrary, has purposes of its *own* which are set up within itself. It may be a mechanism, but the problems it presents are far more complicated than those of any machine. All machines have been made by organisms, but no organism has ever been made by a machine.

Mechanical as a living organism may seem to be, there are important differences between it and a machine. The successive steps in its development or its activity *do not occur in a rigid* sequence, as one would expect if they were aspects of a mechanical system. Jennings has suggested that the organism proceeds by a sort of 'trial and error,' feeling its way, so to speak, towards an end rather than moving by a direct route. The end is the important thing.

Perhaps the most difficult problem in development to explain is how thousands of genes, each affecting a particular process, show such exquisite co-ordination that a precise embryological programme is carried through to its specific end. How this is done is difficult to imagine. Doubtless we shall learn much more about such developmental problems in the future, but the fact that so little progress has yet been made, in comparison with the vast amount achieved in problems of metabolism and physiology, suggests that our mechanical ideas may be too naïve and that something fundamental may remain to be discovered about biological processes.

Whatever the exact mechanism of organization may turn out to be, the mechanist, as a good evolutionist, will suggest that the capacity for growth regulation is simply another character which has been developed by natural selection from random variations and has reached its present high perfection through a long evolutionary process. Why should not the ability of a cutting to strike root and thus restore a missing part be as much a product of evolution as the common, and very useful, tendency of roots to grow downward and of stems to grow upward? Is not the ability to make a whole larva out of one of the cells of the amphibian embryo, if it is parted from the other, the same kind of useful character as the frog's ability to get its oxygen from air rather than from water? If one is the product of natural selection, why not the other? Furthermore, since the co-ordination of development may be upset in tumours, fasciations and other cases of atypical growth, it can be argued that organization is *not* inherent in life but that the regulatory ability of living things has come about in the past, and is maintained in the present, by the continual action of selection.

The difficulty of this explanation is that it explains too much. Many cases of regenerative ability in nature would be called upon so rarely that they could hardly be affected by selection. One must imagine that plants and animals have suffered an almost infinite variety of mutilations and in great numbers, if selection were to be effective in developing their ability to restore all kinds of structures. Furthermore, this ability is often manifest in cases which could never have arisen in nature but only through manipulation in the laboratory.

The regulatory character of living stuff differs from other organic traits in dealing with a fundamental quality that is present in all life and in the absence of which death ensues. Life *is* regulation. Various higher levels of it may disappear—as in abnormal growth and especially in tissue culture—but the basic physiological regulation must continue. Natural selection may modify and improve these activities but the fundamental regulatory quality is that of life itself. In discussing its origin we are discussing the origin of life, not of a particular trait acquired through natural selection.

There is another objection to the idea that selection is responsible for developmental regulations. Their chief visible expressions are the very precise *forms* which the structures of animals and plants display. Some of these are of survival value but many, such as the particular patterns of protozoa and diatoms, or the precise wing venation by which each of hundreds of species of closely related flies can be distinguished, are almost certainly not 'life-and-death' characters. Uexküll and others have emphasized this idea and regard organic form as essentially an independent aspect of an organism, parallel with its matter and energy. Form, they say, is not something developed by natural selection but is inherent in any living thing. This goes a long way towards the old 'idealistic' morphology of Goethe and Cuvier. Indeed, the concept of organization as something independent of the inner and outer environment implies that form must be a basic characteristic of all living things.

The early conclusion, reached in the first flush of enthusiasm over Darwin's great hypothesis, that *all* traits of an organism have resulted by means of natural selection and are therefore of value to it, is now recognized as too general. There are many characteristics of living things which, on any sensible appraisal, cannot be thought of as useful for survival. They are rather the expression of innate co-ordinated tendencies in the living stuff of the organism itself, which thus cannot be regarded as an empty or neutral thing but rather as a system possessed of a directiveness of its own. This is a conclusion of much significance for the suggestion which will here be developed as to the nature of life and mind.

If one is unconvinced by the suggestion that the regulatory behaviour of protoplasm originated and has been perfected by natural selection, what has one to offer that is more satisfactory? The weakness of all anti-mechanistic theories lies in their inability to suggest a reasonable alternative. Analysis, the great tool of science, enables us to break the organism down into smaller and smaller pieces, as a machine might be broken down, but to see how it is put together and works is a problem of synthesis, and much more difficult.

Driesch faced the problem squarely and concluded that although chemical and physical factors of the usual sort are operative throughout most of the processes of organism, no mechanism by itself can produce an equipotential system. 'Something new and elemental,' says Driesch, 'must always be introduced whenever what is known of other elemental facts is proved to be unable to explain the facts in a new field of investigation.'[1] This new factor here Driesch called *entelechy*, something which, from its etymology, 'bears the end in itself.' He does not regard this factor as measurable or as a form of energy, and therefore assumes that it does not violate the principles of thermodynamics. It cannot change the chemical nature of the elementary constituents of the system nor cause chemical reactions which are unknown in the inorganic world. What it does do, says Driesch, is to *suspend* temporarily certain of the reactions which would take place naturally without it. It may later set these free to operate again. Among other examples, he says, 'It is probably in the formation of so-called enzymes or ferments that entelechy manifests itself, not really creating them, but permitting them to appear, whenever required, on the foundation of an enormous amount of possibilities. The ferments, when once in existence, of course, do their work along purely chemical lines.' Thus entelechy is able to regulate development. All life is under its control, which is passed on from generation to generation. Such is Driesch's philosophy of vitalism. The facts to which he appealed are unchallenged, and refusal to accept his conclusions is based on a disbelief in his philosophy rather than on doubt of his results. This hypothesis of Driesch has never commanded very wide assent, however. If one has difficulty in imagining how a living organism operates, he will be equally at a loss in picturing the method of action of entelechy.

Some biologists are in agreement with Driesch. Reinke[2] went part way. He postulated what he called 'dominants,' independ-

[1] Hans Driesch, *The Science and Philosophy of the Organism* (London: A. & C. Black, Ltd., 1908), p. 105.

[2] Johannes Reinke, *Grundlagen einer Biodynamik* (Berlin: Gebrüder Borntraeger, 1922).

ent formative powers, comparable in their operation to intelligence. Others are still closer to Driesch's position, especially in his conception that entelechy (or the 'psychoid') underlies psychical processes. Ralph Lillie in discussing the problem says: 'This psychical factor offsets the physical tendency towards uniformity and dissipation; and under its directive influence, essentially teleological, the routine physicochemical processes are guided and co-ordinated in such a way as to build up and maintain the special biological organization—which in the purely physical sense is so completely 'improbable.' . . . The teleological guidance of a physical sequence requires only an occasional directive intervention; there is no need of a continually active psychical influence which is at variance with the laws of physics.'[1]

Other suggestions as to the origin of biological organization are even more vague and metaphysical. Smuts interprets much of biology through his concept of holism. 'Both matter and life,' he says, 'consist of unit structures which we call bodies or organisms. This character of "wholeness" meets us everywhere and points to something fundamental in the universe. . . . Wholes are not mere artificial constructions of thought; they point to something real in the universe, and Holism is a real operative factor, a *vera causa*.'[2]

Agar, however, does not believe that the whole is of paramount importance in organization but that the parts—cells and larger systems—are in a measure independent and it is their interaction which forms the organism. 'Any theory of the living organism,' he says, 'must take account of the fact that its unity, or wholeness, is not primary, in the sense that it must persist from egg to adult. For the organism can be synthesized out of separate living parts. . . . The cells of the body do not, we may be sure, direct their activities to maintaining the integrity of the functioning body. Natural selection has brought it about that their activities are such as to have this consequence, not striven for by the cells themselves.' And again, 'We

[1] Ralph Lillie, *General Biology and Philosophy of Organism* (Chicago. University of Chicago Press, 1945), pp. 204, 205.

[2] J. C. Smuts, *Holism and Evolution* (New York: The Macmillan Company, 1926), p. 86.

cannot deal with the whole embryo, as the organism concerned, after the point where it has begun to differentiate into autonomously developing parts. From this point onwards we must consider these parts as the behaving organisms, directing their behaviour to their own private, limited, hormic goals, which cannot include the production of the functioning adult organism as a whole.'[1]

Agar's position is evidently based on the old and now somewhat discredited implication of the cell theory, that the cell is the ultimate biological individual and the organism simply a congeries of cells. There are organisms with a considerable degree of complexity in which there are no cell boundaries. The view held by many biologists today is that the division of the body into cells has resulted from the evident advantage this confers in segregating physiological processes from each other and thus making differentiation possible. There is, of course, a good deal of what is called self-differentiation in embryology, each part or system developing to some extent independently of the rest, but the role of the organic whole—the invariable goal of development—is so important that any suggestion that it can arise and maintain itself out of a mass of semi-independent and competing parts and processes is difficult to accept. An organism is indeed a federal union of lower units but one where there is a strong central government to which, it seems, all powers are reserved which are not specially delegated to the parts. The entire, integrated protoplasmic body is the real individual.

A sound philosophy of organism is hard to construct. The vitalist sees the problem but tries to solve it by putting a little agent inside the living system to regulate and control it and thus introduces a variable which seems to be capricious and unrelated to the rest of nature. The mechanist meets the issue simply by ignoring the significance of self-regulation and assuming that a series of physical and chemical changes will organize themselves. The organicist recognizes the basic importance of organization in biology and feels that somehow it is inherent in all of life but he can do little more than speculate, rather blindly, as to what its origin may be.

[1] W. E. Agar, *op. cit.*, pp. 34, 41, 187.

This is a disappointing situation for both biology and philosophy and ultimately for our understanding of man's nature. It calls, I think, for a re-examination of the whole matter, objectively and without preconception. The *fact* of organization is indisputable. To account for it may require some new principles, the discovery of laws in nature that are yet unknown. It would be strange if all the basic principles of the universe had already been found and the only thing science had left before it was filling in details by discovering new facts. The uniformity of nature is a magnificent conception and underlies not only science but our faith in the universe; but there is a range of what seem to be exceptions to it which in time may be brought under a wider uniformity than we now recognize. Among these new and unexpected principles there may be one which will embrace these phenomena of life, now so difficult to reconcile with inanimate nature. Such an eminent physicist as Schrödinger has cautioned us not to be discouraged by the difficulty of interpreting life by the ordinary laws of physics. We must be prepared to find a new type of physical law prevailing in it. Max Delbrück agrees. Says he, 'Instead of aiming from the molecular physics end at the whole of the phenomena exhibited by the living cell, we now expect to find natural limits to this approach, and thereby implicitly new virgin territories in which laws may hold which involve new concepts and which are only loosely related to those of physics.' [1]

That this is likely is suggested by the fact that in recent years a number of new scientific principles have been discovered and are now accepted which not long ago would have been regarded as violating the inflexible laws of nature. Notable among these are the conceptions that the atom is not solid but mostly empty space; that matter is in essence simply electrical energy, arranged in atomic systems; that energy is transmitted not in a continuous stream but in tiny packages or quanta; and that space is limitless but not infinite, for it is curved. Some cosmogonists are suggesting other equally iconoclastic ideas, such as the creation of matter out of nothing. Parapsychology has presented evidence for telepathy and other forms of extra-sensory

[1] Max Delbrück, *A Physicist Looks at Biology*, Trans. Connecticut Acad. Arts and Sciences 38:73—190, 1949.

perception. This, like vitalism, has been largely discounted by scientists since it is outside the realm of natural law as now understood.

The thesis presented in these pages is based on the fact of *self-regulation to goals* in living things and will prove useful regardless of how this regulation is related to physical processes. If we are finally to understand the nature of man, however, it will be necessary to discover the way in which this goal-seeking quality of life arises.

The nub of this problem lies in protoplasm itself, in that very stuff which is the physical basis of life. The fact that life has a physical basis, with characteristics much the same from one end of the animate world to the other, was a biological discovery of the nineteenth century that outranks in importance a recognition that the cell is the unit of structure and function, and a demonstration that living things have progressed through processes of evolutionary change. In protoplasm—this proteinaceous, semi-liquid, flowing, formless stuff—*all* problems of life come to focus. Protoplasm is not a *substance* but a *system*. It is deceptively simple in appearance but evidently has a most exquisitely complex physical structure and chemical composition, for out of it emerge the vast and varied phenomena of living organisms. It is in protoplasm that the basis for regulation and organization is established. On this converge the problems of life.

Our task, therefore, is to try to see how there can be set up in this fluid and unstable medium those steady goals to which growth and behaviour so invariably conform, and what these goals may be. We must regard protoplasm, I think, as possessing *a pattern which so regulates the course of the changes that go on within it that a specific form or activity tends to result.* This pattern is the 'purpose' which leads to the achievement of the 'goal'—the form or activity produced. What the character of such a pattern is we do not know. It may be based on a liquid-crystalline state in protoplasm, or have a bioelectrical foundation, or result from a specific chemical structure, or it may even have an autonomy of its own. That it should *steadily* guide protoplasmic activity seems more reasonable than that something should intrude itself only at certain critical times, as Driesch's entelechy is supposed to do.

In some way the free and creative processes of life, about which we shall have much to say and which are such an essential part of mind, arise by *changes in this pattern* which affect form and deed. The problem is therefore to make possible a certain spontaneity and internal directiveness in it without violating the presumably inviolable laws of thermodynamics and the conservation of energy. This is the rock on which vitalistic theory has always broken up. Can a non-energetic, non-material factor guide the course of protoplasmic change without transgressing physical law? This has recently been discussed in some detail from an engineering point of view by Reginald Kapp.[1] He calls attention (particularly in its application to mental processes) to the difference between *causation* and *control*. The former follows the uniformities of physical determinism in a strict sequence of events so that the organism can be interpreted as a physico-chemical mechanism. At one critical region, however (which I have called the pattern), the varied factors from the outside world that impinge on the organism are so manipulated that reaction is regulated and controlled without the expenditure of energy.

Some such scheme as this, is seems to me, must be postulated for protoplasm, or the most distinctive qualities of life can have no basis. If the protoplasmic system is a purely automatic one, then indeed a living thing is nothing more than a physical machine. If we are to find some way by which creative and, if you will, spiritual forces are to be channelled into life, it is at *this central point of the regulatory activity of protoplasm* that means for it must be sought. If there are specific laws of life, as distinct from those that operate in lifeless matter, it is here that they are to be discovered.

Such a relation between the material and the immaterial aspects of life will certainly not commend itself to most physiologists or psychologists, today, but the impasse in biological philosophy which, despite many soothing and optimistic words, still resists solution requires some bold attempts to break it

[1] Reginald O. Kapp, *Mind, Life and Body* (London: Constable & Company, Ltd., 1951).

down. To this end it may be worth while to push our present thesis still further. If mind—conscious mind as we know it, with all the connotations that this brings—is simply an exalted extension, inwardly experienced, of protoplasmic self-regulation, then whatever we can discover about the mind of man should be applicable, though in much simplified form, not only to our knowledge of such lower levels of mind as instincts but even to the problems of bodily development itself. However superior man may be to all other living things, our argument is that he is in essence like them. Any phenomenon, such as mind or life itself, should be studied not only in its simplest manifestations but in its most complex ones. In man we possess, as it were, a very highly magnified and developed image of the life of a simple organism. And in man we have the great advantage of being *inside* the organism we are studying so that we can perceive by direct experience much that would be impossible to learn from the outside. Although such introspection, because of the variation in its reports and the ease with which it is affected by external conditions, is distrusted by psychologists as a reliable avenue to truth, its evidence certainly cannot be disregarded. Can it give us any information, at this high level, about the phenomenon of purposive regulation which is so evident in bodily development?

If there is any fixed conviction that comes from this inner mental experience, it is that we are *not* machines; that our conscious purposes—identical in origin, it is here maintained, with developmental goals—are chiefly determined by ourselves and not by outside factors. The future seems not to be fixed and predetermined, as it would be if under immutable law, but full of possibility and uncertainty. The feelings of anxiety for the future and of regret for the past, often so vividly experienced and seeming to have the very mark of reality upon them, become quite meaningless if there were no possibilities before us of a choice in what we do. Certainly indeterminism and freedom of choice are so firmly established by experience that man never would have come to doubt them seriously save that he found such freedom to conflict with the presumed inevitability of events under natural law. This problem of man's freedom is a difficult one and we shall return to it later, but certainly the verdict of inner feeling is opposed to an interpretation of purposive action as the machine-like behaviour of a helpless puppet.

Quite apart from this conclusion it is difficult—perhaps impossible—to imagine a human machine which could contemplate itself, and contemplate the inevitability of its own self-contemplation. The mechanistic hypothesis as applied to man gets itself tangled in such a web of paradoxes and contradictions that to accept it requires a degree of metaphysical agility most of us do not possess. Common sense cries out against it, and although we have ample warning today that common sense is often fallible, the practical fact is that we cannot think and act as if we were machines. The very conduct of scientific research presupposes that there are many courses among which we can choose and find our way, many purposes that we can entertain. The empirical basis of this conclusion is as firm as that of any the scientist can reach.

If this is so, and if our purposes and choices are simply protoplasmic goal-seeking writ large, we are bound to conclude, I think, that even the lowly goal-seeking of development has an element in it that is not mechanical, in the ordinary sense of that term. What its nature is remains the final problem. No one can doubt that physical processes are involved in the activities of protoplasm, but just what this added quality may be, this organizing, patterning factor, is still unknown. To identify it with our own conscious purposiveness places this greatest of biological problems on a basis quite different from that which the physical sciences alone can build for it.

What attitude, then, should a biological philosophy take in this great problem of organic regulation? Certainly one of complete open-mindedness and willingness to consider any idea which may be useful. We may have to change radically our conceptions of matter and the mechanisms it makes. The secret of life, organization, purpose, may be lurking in material stuff. Says Paul Sabine, a physicist: 'Unless we assume some miraculous transformation to occur in matter when food is assimilated and becomes the living tissue of an organism, we are compelled to ascribe to the ultimate stuff of the material world potentialities that are beyond the physical and chemical properties of inert matter.'[1] We should heed again the eloquent words of Smuts: 'If we believe that life and mind come from matter, if they are evolved from matter, if matter holds the promise, the dread potencies of life and mind, it can for us no longer be the old matter of the materialists or the physicists. The acceptance of the view for which the materialists fought so hard means in effect a complete transformation of the simple situation which they envisaged. Matter discloses a great secret; in the act of giving birth to life or mind it shows itself in an entirely unsuspected character, and it can never be the old matter again. . . . If Evolution is accepted, and life and mind are developments in and from the physical order, they are in that order, and it becomes impossible to continue to envisage the physical order as purely mechanical, as one in which they have no part or lot, in which they are no real factors and from which they should be logically excluded.'[2]

A certain humility, a certain sceptical or agnostic attitude toward the orthodoxies and the vested ideological interests of biology and the other sciences we should cultivate, rather than the inflexible certainties that now so often find expression. When science forgets itself it can be as dogmatic as authoritarian religion ever was, and in this state of mind it will never be able to reconcile the divergent opinions about life and mind and man that now so seriously divide us.

Here, rooted in living stuff itself, is the fundamental problem of what man's nature is. Not until we come to understand the basic quality of life—its self-regulating purposiveness—can we hope to know the basic quality of man. This is what makes morphogenesis—the developmental side of biology—of such unique importance, for if we could learn the precise control by which a plant or an animal grows from a simple bit of living stuff into an adult with a very specific form and structure, we should then be able to decide what is the relation of purpose and mind to matter.

An advantage of the concept that purposiveness runs through all life is that we can read the series from top to bottom as well

[1] Paul E. Sabine, *Atoms, Men and God* (New York: Philosophical Library, 1953), p. 82.
[2] J. C. Smuts, *op. cit.*, p. 10.

as from bottom to top, and gain valuable knowledge about life from studying man. Evolution, to be sure, is a one-way street and has moved from creatures relatively simple to far more complex ones. What gives it significance to us, however, is that it has shown the wide possibilities that are latent in life. We should remember Lloyd Morgan's conception of 'emergent evolution' which maintains that as evolution progresses new traits and properties emerge which are radically different from anything that had gone before. Thus life may be an emergent from lifeless matter, and the human attributes of mind, reason and spirit may have appeared successively, as living things reached higher evolutionary levels. These would never be suspected from a knowledge of viruses and protozoa. Is it not probable that man still has vast potentialities within himself which at present he has not developed but which will emerge as he grows into something higher? Life continually aspires to higher goals.

In developing the thesis which has here been briefly outlined, it will be useful to examine, in the light of it, various aspects of man's nature and some of the problems about him that so vitally concern us all today.

6

Body

IN THE study of man's nature that we have undertaken, his body is the obvious place for first consideration. Indeed, whether a man is really anything *but* his body is at least arguable. His body is the obvious, tangible part of him; mortal, to be sure, but comfortably material. The first conclusion—and perhaps the most important one—to be drawn from the hypothesis which here has been presented is that this body and its immaterial accompaniment called mind, so often thought of as different and irreconcilable in their qualities, are actually two aspects of the same thing since each is the expression of a basic formativeness or goal-seeking in protoplasm. In the body this expresses itself as a dynamic physical system, developed from a germinal 'idea,' if you will, in the egg. In the mind it expresses itself in the regulatory purposiveness of the activities of this body. Underneath both, however, is the same teleological quality that all life shows.

This conception puts man's physical part in a very different position, philosophically, from that which it has often occupied. It regards the body in a sense as a product of the mind, its first-born, the earliest expression of life's formative activity. The processes which give rise to it are really as much behavioural processes as are those of mind, for the body results from the developmental activities of protoplasm which incorporate matter into a self-regulating living system, and mind acts on this same matter by regulating its behaviour. The 'psycho-physical parallelism,' so long and earnestly debated, in this sense is not a parallelism but an identity. This idea is not new but to the orthodoxies of our day it is still a startling one.

To many whom science has emancipated from the old ideas, a belief that mind, immaterial and thus presumably ineffective, can influence the activity of the body seems very dubious. Body may influence mind, of course, since it is the dominant member of the partnership, but mind can hardly influence the body. Medicine and psychiatry have learned much about these matters in recent years, however, and the very term 'psychosomatic' is the sign of a new attitude. From earliest times religion and the healing art were closely identified. The tribal priest was a medicine man, and this relationship persisted in more refined religions. 'Who forgiveth all thine iniquities,' said the Psalmist, 'who healeth all thy diseases.' In their interest in the spiritual teaching of Jesus, Christians sometimes forget how concerned He was with healing the sick, and His insistence on the curative power of faith. It was as a healer that He first attracted men

Through the ages this healing power of mind, usually in relation to some aspect of religion, was recognized and practised. Numberless examples of it could be cited from the history of Christianity and other great religions. In the days of our emancipation from mysticism and all forms of superstition, however, 'faith cures' and miracles of healing were looked upon with grave suspicion and even the church often lost interest in that aspect of religion which was concerned with health and physical welfare. Here and there a clergyman became concerned with the matter and tried to develop techniques for it. Only in Christian Science on the Protestant side, however, was a strong effort made to relate faith to healing. Meanwhile so many of the cures reported by 'healers' were found not to be permanent, and so permeated was the whole field by quackery and charlatanism, that many rejected the idea of an important curative relation of mind to body. Through it all, the Roman Catholic Church has persistently reported miraculous cures wrought by divine aid. It rejects many claims as unfounded and carefully sifts out a few it finally accepts as true miracles. To read of cures reported by such a man of science as Alexis Carrell, or in the account by Ruth Cranston of those at the shrine at Lourdes, is impressive. To all this, objectors can offer only sheer scepticism and disbelief, not so much because the evidence for such cures is insufficient but because they seem, *a priori*, impossible since they violate the uniformity of nature, that rock on which all science must be founded.

Despite this, the conviction has been growing in recent years that there *is* a far closer relation between mind and body than was formerly thought possible; not a connection that is necessarily miraculous at all, but one in which a mental attitude has a profound effect on bodily welfare. This can heal but it can also grievously disrupt, as psychiatry so often finds. The curative relation begins in simple cases where a firm faith in a doctor's skill has brought recovery, and reaches up to the long and growing list of instances in which a mental or spiritual factor has exerted a healing influence where ordinary medical techniques were powerless to do so. Physicians are coming to realize that the *whole* individual, mental as well as physical, is concerned in health and must be involved in any treatment, and that to minister to single organs or symptoms by themselves is not enough. This is a significant change in the medical point of view.

Religion is again becoming interested in the health of the body as an aid to the health of the soul. The National Council of the Churches of Christ has set up a commission on Religion and Health which is studying a host of cases where 'the ministry of faith, co-operating with that of scientific medicine, is effective in producing healing.' Thus there is increasing recognition, not only in the public mind but in medicine and the church, that mental and spiritual influences are powerful factors in health and disease. This conviction is bound to have an effect on man's physical welfare, for it sees him as a *whole*—body, mind and spirit—and not as a complex of separate parts.

The theory as to the relation between the physical and psychical elements in man which we have been discussing makes this relationship more understandable and indeed something to be expected, for both come from the same source and are simply different expressions of the goal-seeking and co-ordinating power of living stuff. An *organism* is a mind-and-body unity. The problem lies not in one quality of it or another but in the basic and still unknown factors which organize its parts into a whole. This is the problem of life itself.

Why, we may ask, if there is this powerful tendency to guide the development of an organism to a perfect whole and to direct its behaviour, first by instinct and then by rational means, into a favourable course should there ever be sickness or disease or even death? We shall return later to the difficult problem which is involved in the ancient question of sin and evil; but here it should be pointed out that the organizing, goal-seeking tendency of life is by no means infallible. Life often makes mistakes. Unlike a rigid mechanism, its course is not inevitable. It proceeds, so to speak, by trial and error. It is exploring, creative, and the penalty for this is that it can err. Some mistakes are embryological (leading to abnormal structure), some physiological (leading to disease), some mental (leading to error) and some moral (leading to wrong-doing), but they are all involved with the purposive tendency of life.

The suggested identity, at least in origin, of mind and body as opposed to their traditional separateness has important implications beyond those for medicine and health. For many men of faith, and for much of the older Christian theology, the body is something to be put down, kept under or at least in firm control. Man's chief dangers are the world, the *flesh* and the devil. Carnal pleasure, the delights of sense and bodily enjoyment, are frowned on in stricter sects as leading to sensuality and sin. The body, so to speak, is a necessary evil, our temporary house but soon to be outgrown and cast off, as is the way of all flesh. To gratify its desires is a constant temptation and tends to draw us away from our real goal, the eternal life of the spirit. Especially in the Puritan tradition the pleasures of sense are suspect. Much of poetry, music, and the arts were at one time disparaged by it as likely to distract man from his concern with the Kingdom, and no small part of the strength of this stern creed came from its attempt to govern the body and its cravings.

The ascetic life, in Christianity and other religions, goes further still and advocates not only withdrawal from the world and its bodily temptations but a direct assault upon these pleasures by mortifying the flesh through self-denial and harsh discipline. Some of the world's great mystics have come by the ascetic path. But we may ask ourselves, if the physical and the psychical are a unity, why should we despise the one and honour only the other? Why does not the body claim our loyalty and affection as does the spirit? Both come from life, so should not both be served? This was the philosophy of the

Greeks, who delighted in the joys of the flesh and the senses but reached far beyond them to the realm of beauty and higher human values. The Christian suspicion of the body did not originate in Jesus himself who said little about it and whose mission was that men might have life and have it more abundantly. It stemmed chiefly from St. Paul and the early church fathers under whose teachings, during Christian history, life for many lost much of its flavour and richness. Certainly too great a concern with purely physical pleasures can distract us from others that are higher and more permanently satisfying. The lesson our generation needs to learn is that delights that come from music and art and good reading, from love of nature and the exercises of religion are more to be desired than mere tickling of the senses. To rule out purely physical satisfaction, however, is to live a stunted life, surely a poor preparation for the fullness of an eternal one. An occasional withdrawal from the rush and clamour of the world is a healing experience; but a life of permanent asceticism, abundant though its opportunities are for contemplation, seems essentially selfish for it is an attempt to save one's soul by abandoning some, at least, of one's obligations to his fellows.

My contention is not that we should stress the senses as against the spirit but that the body, a necessary portion of our being, should contribute its rightful share not only to our enjoyment but to the development of the higher levels of life. It is through the senses that we not only meet the world around us but are able to perceive in it that richness and variety of beauty which is such an important part of the life of the spirit. Man is not an alien in nature, as Existentialism seems to assume, but through the material part of him is anchored firmly in the physical world. He is like an eddy, someone has said, in the stream of matter and energy which is continually flowing through him.

By means of the bodily senses, too, we often can help our feeble imaginations to grasp truths which unaided they could not perceive. Why, for example, do we delight in broad views, in vistas far extending? Why are we willing to pay more for a house that has a wide prospect before it than for one shut in? Is it not that we thus can grasp physically a little more of space and realize better the inconceivable immensity of it? Without our eyes we could not see the stars. Wide horizons make poets of us all.

Or again, why do we so value ancient things? A chair two centuries old may be physically no different from a modern reproduction but we are willing to pay far more for one than for the other. This quality in man makes possible today a thriving trade in antiques and is so natural that we often do not stop to consider how curious it really is. Does it not come from the fact that a genuine piece of antique furniture assists our feeble imaginations to go back in time and reconstruct more vividly the life of an ancient day than merely reading history could ever do? It helps us grasp the immensity of time. Thus it is, I think, that material objects perceived by our physical senses often aid in our understanding a little better the tremendous concepts of infinity and eternity.

So is it with spiritual things. Though primitive man believed in spirits, he had difficulty in imagining what they were like and therefore erected idols to embody them and make them more real. Let us not be too critical of this humble groping. We should remember that the essence of Christianity is its teaching that the Divine Mystery has been revealed to man through its embodiment in a material human form. Through the ages, furthermore, almost all religions, even the highest of them, have depended on images, paintings, stained glass, noble architecture and impressive liturgies to help nourish and elevate man's spirit. These are appeals to the physical senses, to be sure, but only the sterner faiths deny their efficacy. Physical symbols of all sorts have had a great share in the cultivation of the spirit. Literature depends on written words which are but symbols of ideas. Art itself is the use of paint and marble and harmonious sounds to unveil the beautiful to man.

The body is the window from which the soul looks out at the universe and employs material objects to satisfy its spiritual hunger. It is no alien thing but a colleague of the spirit, manifest in matter and thus able to transmute material values into spiritual ones. We should not pay exclusive homage to it, even to the highest of the delights it offers, and we should accept the possibility that the organizing force which built the body from dead matter may exist when this matter is lifeless once again; but neither should we seek to abase and mortify the body, for it can serve us well. It long has been the model for the greatest of painters and sculptors, and poets have sung its praises.

> Let us not always say,
> 'Spite of this flesh today
> I strove, made head, gained ground upon the whole!'
> As the bird wings and sings,
> Let us cry, 'All good things
> Are ours, nor soul helps flesh more, now, than flesh helps soul!'[1]

One of the god-like attributes of man is this material house he now inhabits. To the Greeks, indeed, he was so like a god in the beauty of his body that in portraying the appearance of their deities, Greek sculptors gave them human form.

> The Olympic Hermes [says Edith Hamilton] is a perfectly beautiful human being, no more, no less. Every detail of his body was shaped from a consummate knowledge of actual bodies. Nothing is added to mark his deity, no aureole around his head, no mystic staff, no hint that here is he who guides the soul to death. The significance of the statue to the Greek artist, the mark of the divinity, was its beauty, only that. His art had taken form within him as he walked the streets, watched the games, noted perpetually the people he lived among. To him what he saw in those human beings was enough for all his art; he had never an impulse to fashion something different, something truer than this truth of nature. In his eyes the Word had become flesh; he made his image of the eternal what men could be.[2]

[1] Robert Browning, 'Rabbi Ben Ezra.'
[2] Edith Hamilton, *The Greek Way* (New York: W. W. Norton & Company, 1942), p. 65.

7

Mind

ESSENTIAL though his material body is, we should all agree, I think, that the most important part of man's nature, the thing that sets him off from the beasts and gives him those high possibilities which as yet he has only begun to realize, is the immaterial portion of him, his *mind*. Here he seems god-like indeed. Various aspects of this psychical quality may be distinguished as different in some degree from one another—the mind in general, the soul, the will, the spirit. In the following pages these will be discussed in the light of the hypothesis briefly sketched already, that man's nature has a basic unity in the self-regulatory character of life.

Mind is a term somewhat outmoded in psychology today and even the existence of mind has been called in question. To the extreme behaviourist it is simply the sum of an individual's acts, and these are interpreted as due to physiological factors and not to such a curious and old-fashioned immateriality as mind. For much of psychology, mind tends to be little more than a convenient term for a group of phenomena that seem to ride along on the crest of the physical activities of the brain without any real autonomy of their own. It is significant that the psychology of today more commonly calls itself the science of behaviour than the science of mind.

'Mind' in the old sense was thought of as that which governs bodily acts and comprises conscious sensations, emotions and thoughts; something limited to man or perhaps the very highest of the beasts and which therefore must have arisen and been perfected during the process of organic evolution. This raises the difficult question of just where and how mind could have been interjected into the body, not only in evolution but in the embryological development of an individual, as well.

In the hypothesis that has been considered here mind has no limited extent like this but is present, at least as to its rudiments, in *all* life. It is rooted in purpose, which is another name for the self-regulatory and goal-seeking character of protoplasm. This concept is so wide that it will cover much that would commonly not be regarded as 'mind' at all, but it has the advantage of avoiding both the perplexing division of man's nature into two distinct parts, one physical and the other mental, and the problem of how mind emerges in evolution and in development. It rescues mind from the limbo of unreal things by relating it closely to a definite process in the body, though without commitment as to what this relationship implies.

Let us then for purposes of this discussion, define mind as *whatever directs the development and activity of an organism towards goals set up within its living stuff*. This is a definition which, I am sure, few psychologists would recognize as having much relation to 'mind' at all. It puts mind into the very simplest of the animals and even into plants. It begs the question of precisely *what* mind is, for by stretching the definition a little further, the sum of the physical factors which maintain and operate any automatic self-regulating machine could be thought of as a mind. In the definition here presented, the essence of what mind 'really' is hangs upon what the regulatory factors are that govern the activity of an organism. This is the final question for both biology and psychology.

Conscious purpose may be interpreted as *the inner experience or awareness of the operation of these regulatory factors.* Consciousness which pre-empts the discussion of much of psychology and philosophy today, is certainly of great importance and is so closely associated with mind that it is regarded by some as a necessary attribute of mental life. The *awareness* of regulation, however, involves a different problem from that of regulation itself. How far down in the scale of beings consciousness extends has often been debated, and as in so many other human traits it is hard to find the place where it first appears. As mere sensation, the awareness of a stimulus from outside the organism, it may be co-extensive in evolution with protoplasmic reactivity itself and thus with all life. In evolution, as the complexities of stimulus and response become much greater, the time arrived, probably with the evolution of the highest mammals, when the individual began to be aware of more than his sensations alone and perceived himself and his own life in relation to the rest of nature. He stepped outside himself, as it were, and became an observer of the whole scene. The 'I' recognized the 'me.' Here consciousness, in the commonly accepted sense, was born. Consciousness is indeed mysterious, and in some way it is closely tied to life; but we need not here discuss the various problems of metaphysics and epistemology which a consideration of it is bound to involve.

Unlike life, this vivid 'focal' consciousness is not continuous but lapses, in sleep and under the influence of narcotics, into unconsciousness. The line between the two is hard to draw, for one merges insensibly into the other. Psychologists recognize a wide range of the unconscious which has, so to speak, a submerged life of its own where memories are stored up and mental work is often done without our awareness of it at all. From the unconscious there surge up into consciousness a host of feelings, impulses and other mental influences which greatly affect all that we think and do. Their importance in mental health is well recognized today. The basement of the mind, this unconscious layer of it, is closer to the physiological and the protoplasmic than focal consciousness is and here, doubtless, the regulatory and purposive character of living stuff is more directly manifest. Unconscious desires and purposes powerfully affect conscious ones. My point is that this purposiveness is everywhere in the life of the mind, whether through consciousness we are aware of it or not. Mind, rooted in protoplasmic purposiveness, is continuous. Mind *is* the organism, in a sense. Conscious mind, the highest expression of mental life, emerges from this basic purposiveness with various degrees of intensity, from bare awareness to vivid 'focal' consciousness.

The underlying question in all this is the relation between mind and its physical basis in the organism. That a close relation does exist cannot be doubted, for mind is manifest only in a living body. Aristotle and the ancients thought the heart was the seat of consciousness and of all mental life, and other erroneous ideas about the relation of mind to body were proposed. Mind in the common use of the term seems now almost exclusively based on one kind of tissue and one set of organs, the nervous system. Nerves are gathered into cords or trunks, of which the spinal cord is the most important. At the head end of this cord is a huge mass of nerve cells, the brain, enclosed within the bony shelter of the skull. In the ascending evolutionary series the intelligence of an animal, or at least the complexity of its psychical processes, increases in proportion to the size of its brain, and there is now no doubt that the brain is the seat of what we call the mind.

The brain is an organ of the most remarkable complexity. It consists of about ten billion nerve cells. These connect with others extending out as long and very thin fibres to a particular sense organ from which stimuli come, or to particular muscles, glands or other structures the activity of which is to be controlled. The impulses which a nerve fibre conducts are evidently electrical ones and move with high speed along its surface sheath at from three to ten per second, each lasting about a 1/10,000 of a second. The contact of each fibre with others may be made or broken at contact points or synapses, and the whole enormously complicated system suggests a telephone exchange. Space is lacking here to describe the fantastic elaboration of the brain, and much about it, indeed, is still unknown. It is in this organ that the contact of mind with matter evidently occurs. Here is the centre of our problem.

Much of the work of the nervous system is involuntary and does not involve conscious mind at all. By reflex action a stimulus automatically calls forth a response. Such is the knee-jerk or the blinking of the eyes. Many reflexes can be conditioned by gradually substituting one stimulus for another. The simple and inevitable reflex often becomes more varied as several act together and interfere with one another. Some activities, such as breathing or the beating of the heart, are essentially reflex acts which are self-stimulating. Habits are acts originally voluntary but which finally become essentially reflex or at least go on without voluntary control. By extending this conception more widely, the whole of mental life has often been interpreted as reflex or automatic. This idea of man as an automaton was first formally proposed by Descartes in the seventeenth century and, in essence, is still supported by many psychologists.

But the complexities of behaviour can hardly be explained as a series of simple automatic reflexes alone, for behaviour is governed by many factors, notably memories of past events. Much of what a dog does, for example, can be explained as a series of reflexes—his walking, snarling. shaking his wet coat, eating, drinking and so on. These he will do almost as well when the bulk of his brain has been removed. But as Sherrington says, a purely reflex pet could never please us. A dog is far more than this.

That conscious mind is biologically useful is strongly suggested by the fact that it has been developed to its present state during the hurly-burly struggle of evolutionary ascent, when the traits of an organism presumably had to prove their worth. Mind is chiefly concerned with those organic processes which are readily modifiable and it would seem, if common sense and our innate convictions are to be trusted, that it actually does control these processes.

Here we meet the difficulty head-on, for mind is not energy nor any form of energy! How, therefore, can it affect a material system? The plain answer of materialism is that it cannot do so. Sherrington frames its answer thus: 'Thoughts, feelings and so on are not amenable to the energy (matter) concept. They lie outside it. Therefore they lie outside Natural Science. If as you say thoughts are an outcome of the brain we as students using the energy-concept know nothing of it; as followers of natural science we know nothing of any relation between thoughts and the brain, except as a gross correlation in time and space.'[1] If this is so, mind and all its concerns are not part of science at all. Psychology would not be a science. Medicine, so far as it deals with the health of the mind, must be outside the scientific pale; but this surely no psychiatrist would admit. Here is another flat contradition between points of view which seem irreconcilable.

Sherrington's great book is a long and vigorous wrestling with this problem of mind and body, illuminated with all the knowledge of an able scientist and written with clarity and charm.

[1] Sir Charles Sherrington, *Man on His Nature* (New York: The Macmillan Company, 1941), p. 290.

The author, as a convinced mechanist, endeavours to interpret mind in purely scientific terms but always he has trouble in containing it within these limits. It continually bursts out here or there from the bonds in which he seeks to hold it. He is too much of a philosopher to overlook the serious difficulties which confront the man of science who tries to deal with mind, and his conclusion does little more than face this difficulty.

> But though living is analysable and describable by natural science, that associate of living, thought, escapes and remains refractory to natural science. In fact natural science repudiates it as something outside its ken. A radical distinction has therefore arisen between life and mind. The former is an affair of chemistry and physics; the latter escapes chemistry and physics. And yet the occurrence of mind—recognizable finite mind—is confined to a certain particular field of chemistry and physics, namely, that of highly integrated animal lives. 'Thinking,' in this its limited field of occurrence, appears as a phase of living. If, as is practical, we continue to subsume mind under life, we have to distinguish it as an activity of life selectively and uniquely apart from the rest. The psycho-physical difficulty places us in the position of empirics as to much. By ways which may be judged roundabout, we find ourselves at length pragmatically alongside of general common-sense opinion. That may be taken either as sanity or superficiality or perhaps both.[1]

In other words, the problem seems insoluble.

It is just here, I think, that the suggestion we have been developing in earlier chapters may be of service to help to break this impasse. Mind is not energy, but neither is the goal, set up in protoplasm, to which development and behaviour conform. This goal, as discussed in Chapter V, is a pattern in living matter—in simpler cases rather firmly fixed but in higher ones, as in cells of the nervous system, subject to continuous change—which regulates protoplasmic behaviour. It is not energy but simply establishes the pathway along which energy will flow.

[1] *Ibid.*, p. 291

How it is set up and maintained we have already considered though to no conclusion. The important fact is that somehow it *is* set up. The protoplasmic pattern, immanent in the egg or in any cell, to which the various stages in development and finally the structure of the mature individual will conform, is the prototype of a purpose, and finally of an idea, that is immanent in the cells of the brain. The function of the brain, and particularly its cerebral cortex, is admittedly to unify and regulate behaviour. The brain is a great clearing-house where stimuli of all sorts are received, both from the sense organs and from sources inside the body and in the brain itself, and where the reaction to these is so directed that it will *tend to maintain the goals*—often changing—*which are set up in the living stuff of the brain cells*. It stands between stimulus and response, and performs, on a grand and complex scale, the same regulatory, purposive function which at lower levels is carried on by a simpler nervous system or, in development, by the more generalized protoplasm of the body cells. Early ideas that the various mental processes are sharply localized in different parts of the brain so that behaviour is, as it were, compartmentalized, are not supported by the evidence today. The whole brain is concerned with most behaviour and with memory, imagination and reasoning power, and it is the whole brain that sits in the driver's seat and steers behaviour along the road to the achievement of goals.

This suggestion, it seems to me, places the whole vexed issue of the mind-body relationship in a somewhat different light and avoids the unhappy dilemma in which men like Sherrington now find themselves. The mind is not energy, to be sure, but it is intimately involved with matter through its relation to the goal-pattern in protoplasm by which energy is directed. Conscious mind is the *experience* of the operation of this pattern. It need not expend energy any more than a thermostat does.

Mind is certainly as open to scientific investigation as any factor that is related to a material configuration, such as a bioelectric field or a similar pattern in space. Thus we reach a conclusion opposite to that of the distinguished physiologist whom we have been quoting when he says that mind is more ghostly than a ghost, something invisible, intangible, and without sensual confirmation. Surely, we might reply, the thing that makes man great and god-like must be far more than such a ghost.

The hypothesis presented here maintains that psychology—the science of mind—deals with the same material as biology though at a much higher level. Both are concerned with patterns in protoplasm to which activity conforms. We must leave unsolved the problem of how these patterns are established, for this is the problem of life itself. When we have reached an answer to it we shall understand not only life but mind and thus many problems of philosophy. All this suggests that biologists should pay more attention to the fundamental fact in their discipline, protoplasmic organization; and that psychologists should explore more fully the relation of the science of mind to the science of life.

That mind indeed is thus primarily regulatory and is an integrating factor in life is plausible, but there is further and most interesting evidence from quite a different source. This comes from the fact that the mind perceives in patterns, figures or *Gestalten* which are more than the sum of individual sensations. Each of the millions of cells in a sense organ, such as the retina of the eye, when stimulated, passes along a stream of impulses by a fixed path which excites corresponding cells in the cortex of the brain. These individual impulses are not received as individual sensations, however, but the mind in some way organizes them and makes of them a form or pattern which is perceived as such. If this were not so we should see and hear nothing but a jumble of messages from the outside world. When the eye looks at a circle, the stream of impulses from the image at the surface of the retina to the brain is not circular but out of the stream the mind in its perception is able to reorganize them as a circle, much as a 'scrambled' telephone message is unscrambled at the other end. In the same way, a series of sounds is perceived not as a jumble of auditory sensations but as a melody. When a circle of a different size or colour is seen, it is still recognized as a circle, and a melody played in different keys is recognized as being the same. Whatever cells in retina or brain are involved, out of their activity—decoded from it, so to speak—comes the same message, the same Gestalt.

The way in which the mind organizes a swarm of apparently unrelated stimuli into a pattern or whole and perceives this whole through a wide range of changes in the stimuli themselves is strikingly similar to the way an animal or plant builds random matter into the organized system which is its body, often in spite of many obstacles. The subordinate relation of the parts (individual stimuli or individual cells) to the whole (the Gestalt or the organism) is much the same in one as in the other. Von Bertalanffy well says that 'the dependence of parts on the whole, as in Driesch's sea-urchin experiment, is not a vitalistic feature, but a general characteristic of Gestalten.'[1] The general fact of biological organization, chief theme of our present discussion, thus expresses itself both in developmental regulation such as Driesch made famous and in the mental ones Gestalt psychology discloses. One is body and the other mind, further evidence that the two are fundamentally alike. There is a morphogenesis in both.

All this emphasizes again what psychology and neurophysiology both conclude, that mind is an *integrating* agent, that it sorts out the vast array of sensations and governs the host of possible responses so that unified and orderly behaviour results. Normal individuals do not switch crazily from one line of action to another, but pursue a continuous course of behaviour. In a system that is in contact with such an immense variety of external things and can make so many responses to them, this is possible only through such an agent as the mind is. It is what preserves our *selves* for us. In various pathological conditions, notably cases of 'split' personality, this unifying process is defective, and many other mental ills result from failures in its operation.

The problem lies not only in the regulation of behaviour so that it maintains a unity, but in what this behaviour actually proves to be, the goals to which it will conform. This problem is one that we meet everywhere. Mind is not only the unifying agent but the directing agent. It not only keeps the individual on one track but it determines where that track will lead. The great question about man's behaviour is why he does one thing rather

[1] L. von Bertalanffy, *Problems of Life* (New York: John Wiley & Sons, 1952), p. 192.

than another. What determines his acts? Since he is a social organism and since his conduct is therefore vital for his fellows and for the successful operation of his society, this question is of the utmost moment. To govern conduct, taboos have been erected, moral codes devised and in civilized communities elaborate bodies of law formulated. To keep alive the organism we call a man is by itself a complex problem of the homeostatic regulation of his physiological processes, but to determine what he *does* with this machine is a far more significant one. Both involve that regulatory, goal-seeking quality of living stuff which so characteristically distinguishes it. Out of this come decisions and directives which are of such pregnant consequence as to lift the problem of man's conduct to a level far above a brute's. One's success or failure may ultimately be due to the environment in which he is born or to the genetic equipment with which he happens to be endowed, but its immediate cause is the strength and the direction of his goals, of the desires that he strives to fulfil. They may make him god-like indeed or cast him down again to the beast from whence he came.

To explain behaviour merely by saying that it is due to an act of mind or of will does not satisfy a psychologist, nor should it. No act merely happens. There must be a motive for it, and motivation is therefore the key problem in behaviour. Just what the origin of motives is, just what determines their character and strength, therefore becomes a matter of great consequence. Most psychologists here adopt the orthodox mechanistic position. This regards motive in a man as essentially like motive in a machine, the actual physical motive power that makes it run. Energy must be guided in its expenditure by the directive mechanism of the brain, which drives the body along a course of conduct. This impulsion is well termed a *drive*, and much of psychology today is concerned with a study of such behavioural drives. Their strength and direction are set by various conditions in body and brain. Hunger is a drive which pushes an organism along the path of action by which it will be satisfied. Sex is another powerful drive, rooted in the ancient urgency for reproduction, without which any species would soon cease to be. There are drives at higher levels than these purely physical ones—drives for sensual gratification, for power, for adventure, for creativeness and many more. By them an organism is pushed along a course in this direction or in that, sometimes strongly, sometimes feebly, but always pushed.

The hypothesis we have been considering here suggests a basis for motives different from this. If goals and purposefulness in their simplest form are what determine the behaviour of the lower organisms, it is reasonable to think they do so in man himself. The actual energy that moves his body, of course, has a physiological source, but the direction in which he goes is determined, I think, not by a mechanism in the ordinary sense of the word—a mechanism like that which drives a car—but by a protoplasmic directiveness which sets up a goal towards which the organism moves. The inner experience of this goal may lead to immediate action to attain it, or this action may be checked and lead simply to a desire for the goal. We perform an act primarily because we *want* to do it. In terms of the old fable, it is the carrot rather than the stick that is more important. Purpose and desire are vital and effective agents in conduct. As we have said so often, this does not mean the sort of teleology in which the future influences the present and 'final' causes are effective; but it does mean that goals set up within our living stuff, presumably in the brain, determine action. Motives are what in common sense we think they are—desires, wants, purposes. When hungry we do not feel as though we were being *driven* towards food but rather that we *want* it very much. Ambition does not seem to be the result of an inner push but rather of a deep and sometimes a consuming desire. To some extent psychology recognizes this in its concept that *tensions* are set up within us and that we are driven to reduce them.

The conception of goal rather than drive as the basis of motivation is much more in harmony with a philosophy that puts the encouragement of high ideals and aspirations as the best means of elevating mankind rather than with one which depends on environment and conditioning and finally on physiological mechanisms to do so. The latter are certainly important, but unless man learns to *want* what is beautiful and good and true, he will never attain these things. The subtlest form of propaganda seeks to modify not what we know but what we wish, and the most effective leadership and guidance are directed at the hearts as well as the intellects of men. The task of education at every level—intellectual, moral and spiritual—is to cultivate a *desire* for knowledge and right conduct and spiritual things. Until men learn to *want* all these, no palliative will lift the world out of its present sorry state. It is the drawing power of the ideas and ideals of great leaders and teachers that has inspired men through the ages, not only for good but often for ill. He who learns to mould men's desires can stir them to great deeds and thus move the world.

Between the philosophy of 'drive' and 'draw' there may not seem at first to be much practical difference, but fundamentally they are as diverse as the poles. Which we support will determine how we live. Few tasks are more important, for the psychologist, sociologist, educator and humanitarian, than a study of the roots of purpose and desire and of means by which they can be moulded. Here, I believe, the concept of goal-seeking as a fundamental property of life will prove its worth.

But mind involves much more than a control of bodily behaviour and of conduct. Even when the body is quiet, a stream of thoughts is 'passing through the mind.' Whence do these come and how are they related to goal-seeking and purposiveness? They are of vital importance, for their highest expression, abstract thought, makes possible man's power to reason which has put the world under his dominion.

In animals a purpose, presumably unconscious or at best semi-conscious, is translated directly into act. There may be several purposes in conflict with one another but the stronger prevails and issues at once into behaviour. In man, however, this is by no means always so. A purpose may express itself not in a physical act but in a mental one—an idea or a thought. Some behaviourists are inclined to think that even these involve bodily acts, almost imperceptible movements of the larynx—'subvocal articulation'—by which unspoken words are framed that embody conscious thoughts. Most mental acts, however, apparently exist in the mind alone. Just as the goals in development or behaviour are constantly changing as a person—or an animal—engages in different sorts of activity or as he moves from one environment to another, so these mental goals are constantly altered. Much of the stimulation may now come not from the outside through organs of sense, but from previous mental acts themselves, so that a stream of thoughts succeed each other. Purposes are thus enriched through memory and imagination.

This may seem a rather naïve and inadequate picture of such a complex thing as thought, but if physical life is rooted in protoplasmic purposiveness it is reasonable to believe that even these more elaborate forms of mental activity are the expressions, much modified, to be sure, of primitive goal-seeking. The physical basis of a thought is presumed to be the same as for a bodily act—a goal set up in living stuff. If we could discover how this is brought about in such a relatively simple process as development, we should be close to knowing how life and mind are rooted in matter.

These higher goals present the difficult problem of how they are established. If man is free and creative (a proposition which we shall shortly attempt to defend) these qualities are manifest primarily in the way he sets up the throng of goals which are his mind, for they determine his thoughts and his deeds. What they are depends in part, of course, on outer circumstance, but this works through his inner directive. He not only keeps his craft on her course, tossed about though she may be by wind and wave, but determines what that course shall be through all its many changes.

The practical importance of all this is that through thought we are able to recognize relationships between things and between ideas. Some of these are very simple. A savage, for example, may see the general fact that seeds put into the ground grow into new plants like the ones from which they came. This requires sharpness of observation and capacity to generalize. He may thus infer that to grow a particular kind of plant that he wishes he must sow seeds from that type of plant. From such simple acts of reasoning it is possible to go on to far more abstract ones and finally to mathematics, logic and philosophy, the relationship between ideas.

That the mind is able to relate observed events to each other, then ideas and finally abstract thoughts may have a deeper meaning than is evident at first. The essence of protoplasmic activity is *organization*. Even at the lowest level, protoplasm makes and maintains complex patterns of relationship not only between the parts of an organism but between the various activities it performs. These patterns are the goals to which it conforms. The fact that protoplasm is concerned primarily with relationships may not be without significance even at the mental level of life. As living stuff effects a precise relationship of the parts of an organism to each other, so the mind—a manifestation of this same organizing capacity—can relate *ideas* to each other. Reason itself may thus be regarded as the highest expression of biological organization! Perhaps this suggestion is not so fantastic as it might first appear.

The concept which is the core of our present discussion, that protoplasmic self-regulation is the basis of purpose and thus, in time, of mind has been able to offer several suggestions about the mind and its activities that are worthy of consideration. They are psychologically unorthodox; but they present a method of approaching this great problem which to a biologist seems of value. It is man's mind that sets him off above the beasts and is the basis of much of the god-like in him. In the study of his true nature which we are here attempting it must occupy a major place.

8
Self

IN MAN and the higher animals the complex activities in nervous system and brain regulate behaviour and thus knit the organism together as a single whole. They make of it an individual. The organizing character of protoplasm, of which this is an example, is manifest at various levels. First it produces individual cells, each an organized system on a small scale. In the simple unicellular animals and plants, this is the individual. At higher evolutionary levels the body consists of a group of cells and finally of very many of them, indeed of many billions in the higher forms.

The discovery over a century ago that the bodies of plants and animals are made up of cells marked a great advance in biology, corresponding in a sense to the discovery of atoms as the units of matter. It made possible a much more exact analysis not only of the structure of living things but of the physiological activities that take place in them.

So dominant was the cell theory at the turn of the century that most biologists believed the cell to be the true living individual and a multicellular body a sort of colonial cell society, its structure and properties the result of interactions among the cells which compose it. This view has now been widely challenged. In many of the simpler plants and animals, to be sure, masses of similar cells may remain attached to one another instead of separating, and thus form true colonies. From these colonies the many-celled individual may have evolved. Almost from the beginning differences appeared within it, the first step in that differentiation among its parts which marks a true individual. Larger and well-integrated organisms followed these semi-colonial types. Here each cell retains its individuality and a certain independence of its own in development and function.

So do many organs, particularly in plants and the less highly specialized animals. All parts of the many-celled mass, however, are physiologically related to each other and to a degree are dependent on each other. The organism is not simply an aggregate of cells but an integrated system. Most biologists would agree today that the true individual is not the cell but the entire many-celled living body. That the cells are distinct from one another is advantageous since this makes possible the separation of different chemical substances and physiological processes among different cells and thus allows for a differentiation and division of functions which would be impossible if substances could move about unhindered by cell boundaries. This advantage was probably important in evolution and may explain the almost invariable occurrence of cellular organization in living things.

The individual organism is the major unit in biology. Living stuff is moulded into individuals by its property of organization. Where this is relatively weak the individual is not sharply limited, as in many plants. Parts of such a diffuse plant system as a group of strawberry plants are almost separate individuals even though still joined together, and certain colonial animals are groups of connected individuals. A higher level of organization results in much more closely integrated individuals, as in any of the higher animals. Throughout all life, protoplasm rarely makes diffuse and formless masses, save in abnormal growth where the organizing force is partly lost, but tends strongly to pull itself together, so to speak, into living units, or organisms. A *single, whole organism* is the goal towards which development proceeds. This wholeness is immanent in all its parts. If these parts lose continuity with each other, each part, as we have seen, tends to restore a whole. This formation of individual organisms is one of the conspicuous qualities of life. It is well described by Jennings. 'Life,' he says, 'does not occur and develop as a single unified phenomenon, as one continuous organism. On the contrary, we find life now occurring and developing in a great number of diverse centres separated in space. These separated centres we call individuals; myself, yourself, our friends, my dog, the sparrow I saw in the tree as I walked in the park. Each one of these individuals is an integrated group of structures, activities and inner experiences. Each has sensations, emotions and mental experiences that are separate from those of the other individuals. . . . Each is a centre of experience, a spectator of the world as well as an actor in it; a centre which retains its separate identity, its "personality" among the changes that occur.'[1] This aspect of living things has been vigorously emphasized by Smuts in his philosophy of holism, in which the concept of the organism as a *whole* in all its aspects is stressed.

To maintain the unity of this individual, in plants and the simplest animals, is a function of the entire mass of living material. The unity shown in bodily development among animals is extended to a unity in behaviour which finally comes under the control of a special group of cells, the nervous system. Each individual tends to have a consistent pattern of action, an individuality in its behaviour as well as in its structure, and this is maintained by the unifying action of the brain.

In man, it is the character and significance of the biological individual which he is that raises the difficult problem of personality, the ego, the self and finally the soul. Just what is this elusive individual unity? Has it a real existence of its own? Does it play a part in the control of behaviour? Are there values connected with it which man should recognize? And, finally, is there any possibility that it can exist apart from the body that it now inhabits? These are questions which concern the very basis of man's nature. It is what he is *as an individual* that is of first importance.

With these problems psychology long has wrestled. It is so easy to assume a self or a soul and by this means to 'explain' what a man does without really accounting for it at all that the modern science of the mind tends to abandon the concept of the self altogether. It regards a human being as an elaborate mechanism in which stimuli elicit the responses we call behaviour. Thought is the result of complex interactions in the brain and need not involve a thinker at all. Man is an 'empty organism' and everything that happens to him is initiated from the outside and never arises actively from within. 'A psy-

[1] H. S. Jennings, *The Universe and Life* (New Haven: Yale University Press, 1933), p. 52.

chology without a soul' is its ideal. 'For two generations,' says Professor Allport, 'psychologists have tried every conceivable way of accounting for the integration, organization and striving of the human person without having recourse to the postulate of a self.'[1] In a vigorous plea for recognition of a self in process of becoming, he states his belief that 'an adequate psychology of becoming cannot be written exclusively in terms of stimulus, emotional excitement, association and response. It requires subjective and inner principles of organization of the sort frequently designated by the terms self or ego. Whether these labels are employed is less important than that the principles they imply be fully admitted in accounting for the development of personality.'[2]

Here the concept of biological organization as a principle can make an important contribution, for it assumes the presence in protoplasm—and thus in the brain—of norms or goals to which the activity of the organism tends to conform. The character of these goals changes in development as one stage succeeds another, and in behaviour as one purpose succeeds another, but the fundamental fact is a *continuity* of goal-seeking. Here is no hypothetical, immaterial self that is running the machine but a constant protoplasmic pattern—perfectly mechanical if you wish to think of it as such—through which matter and energy are channelled to a specific end. No mere series of reflexes could give this result. An organism in its very nature is a *continuous* system. This must be accounted for, whatever theory we build. How can any mere aggregation of material particles preserve the past into the present in the sort of duration we experience and that Bergson has emphasized? One configuration of particles could succeed another but how can there be a recognizable continuity of the *identical whole* unless some means for it is provided beyond the random give and take of reflex action?

A living organism is the sum of its various goals. They bind it together and are the source of the unity it so strikingly displays, a unity not only in space but in time. Every protoplasmic system has a history. This is not a mere repetition of reactions

[1] Gordon W. Allport, *Becoming* (New Haven: Yale University Press, 1955), p. 37.

[2] *Ibid.*, p. 60.

but, as Allport shows, a real *becoming*. In the pulse and stir of time an organism persistently maintains its own identity. 'Like the universe as a whole,' says Bergson, 'like each conscious being taken separately, the organism which lives is a thing that *endures*. Its past, in its entirety, is prolonged into its present and abides there, actual and acting.'[1]

All this is the more remarkable since the organism is a centre of great activity and change. Matter is continually entering and leaving it. By the use of 'tagged' atoms, the course of which in the body can be followed, physiologists are able to measure the rate of this 'turnover.' They find that it varies in different organs but that everywhere a new bit of matter simply takes the place of one that is leaving. The organism is not a rigid unity but a dynamic equilibrium, an 'open system.' As Heraclitus observed, long before the days of modern physiology, 'Man is like a fountain, always the same form but never the same water.' In all this flux of matter and energy the organism stubbornly maintains its own particular individuality. The material that constantly pours through it never makes it into something else. One of the most remarkable facts in biology is this persistence of the individual.

How this dynamic system regulates itself so precisely as to maintain the delicate equilibrium necessary for life is difficult enough to understand, but how at the same time it preserves its own specific character is a harder problem still. Something at the core of it, something that in a man we call his personality, is at the bottom of this continuity. The individual is a knot of potencies, purposes and desires almost impossible to loosen. This is what gives man his unity, maintains his identity, binds him to his past, and makes his future possible. These goals are what we *are*. Not *cogito, ergo sum,* we well might say, but *desidero, ergo sum*.

To this problem genetics can offer an important contribution and one which psychology is sometimes likely to ignore. Each of us has his own specific inherited constitution, the basis of which is a complement of many thousands of individual genes, duplicated in every cell of the body. These are distributed

[1] Henry Bergson, *Creative Evolution*, Tr. by Arthur Mitchell (New York: Henry Holt and Company, 1911), p. 15.

among the twenty-four pairs of chromosomes, each chromosome with its particular genes arranged in linear order somewhat like beads on a string. There are two sets of chromosomes in each cell, one coming from the father and one from the mother. When sex cells are to be formed the chromosomes line up and one from each pair moves into a daughter cell which will finally give rise to a sperm or an egg. Each sex cell thus has only half the chromosomes of a body cell. In this separation, members of the maternal and paternal chromosome sets do not go together into the same sex cell, and genes from the two parents are therefore thoroughly shuffled and recombined in their offspring. Furthermore, a pair of chromosomes may exchange part of their genes, and there may even be exchanges between members of different pairs. As a result of all this in a species like man, which is highly mixed genetically, gene shuffling and scrambling increase the variability. In addition, genes occasionally mutate, thus augmenting the store of genetic differences. When a child is born, the chance of its being genetically a duplicate of any other human being on earth—or even any other who has ever lived—is infinitesimally small.

All this is significant for the problem of the self since it means that every individual has his *own* specific and particular genetic constitution, unlike that of any other person. He is not turned out by a process of biological mass production of similar individuals and interchangeable units, like machines from an assembly line, but every individual, so to speak, is custom built. If each of us is a mechanism, at least he is a highly distinctive one and unlike any other. This fact has always been disturbing to those who look to the environment almost exclusively as the factor by which men are moulded. With its wealth of specific genetic differences the human organism is far from empty. No geneticist believes, of course, that everything we are and do is genetically determined, for both inner and outer factors are important in moulding us. If all men were alike, turned out on the same identical plan, how stale and monotonous life would be, lacking that vigorous contact between diverse personalities which gives it so much zest and flavour. We can hardly imagine friendship or altruism existing in a society of such beings, or at best only the friendship of a beehive or an anthill.

All these facts suggest that there is present in a living organism, and most conspicuously in man, a central integrating factor, the sum of its regulatory activities, and specifically different in each individual, which deserves the name of *self*. How this is related to the ego, the id and other concepts of such a centralizing agency is not important. *Something* is involved in the development and behaviour of an organism that is autonomous, internally controlled and not merely a series of responses to stimulation. Even in the case of Spemann's 'organizer,' a bit of the early embryo which once was thought to control the organization of the whole, it has been shown that the organizing activity resides instead in the living stuff of the entire embryo. The self can react to outside influences but it nevertheless remains a specific, protoplasmic whole and has, if one may say so, a will of its own.

One should study a problem like biological individuality or the self, I think, not only in its simplest expression in lower forms of life but in its highest development, in man. The finest expression of the self is a human *personality*. As Julian Huxley says, 'This primacy of human personality has been, in different ways, a *postulate* both of Christianity and of liberal democracy; but it is a *fact* of evolution. By whatever objective standard we choose to take, properly developed human personalities are the highest products of evolution; they have greater capacities and have reached a higher level of organization than any other parts of the world substance.' [1]

Man's personality seems to be so much more than mere protoplasmic integration that one hesitates to think of the two as basically equivalent, but as good evolutionists we can hardly deny that the complex expressions of selfhood and personality in man are manifestations of something that began as simple biological individuality. We are acquainted with human personality in our friends and have a vivid inner experence of it in ourselves. It is so rich and various and so obviously the fine flowering of the human organism that it seems to be far more

[1] Julian Huxley, *Evolution in Action* (New York: Harper & Brothers, 1953), p. 165.

than merely a series of reflex acts, a give and take between the brain cells and their surroundings.

The very fact of separateness between personalities engenders a sense of individual value that would be impossible in bits of a continuous living mass. The feeling I possess of my own personality as I face my individual responsibilities and destiny has in it the very flavour of reality. It is only in persons that ideas are born, by persons that discoveries are made and poems written. It is only as persons that we can communicate with one another, and this may be the only way in which we can finally make contact with the universe itself. These facts, it seems to me, can tell us something of the phenomenon of biological individuality by direct experience of it at its highest level which no amount of objective study of animals or man could ever give. Life is ultimately understandable only in terms of personality.

Perhaps because personality does not fit into the pattern of modern psychology, that science has rather little to say about it. This is regrettable, and even as laymen we are inclined to agree with Professor Allport when he gently chides his colleagues: 'It is especially in relation to the formation and development of human personality that we need to open doors. For it is precisely here that our ignorance and uncertainty are greatest. Our methods, however well suited to the study of sensory processes, animal research, and pathology, are not fully adequate; and interpretations arising from the exclusive use of these methods are stultifying. Some theories of becoming are based largely upon behaviour of sick and anxious people or upon the antics of captive and desperate rats. Fewer theories have derived from the study of healthy human beings, those who strive not so much to preserve life as to make it worth living. Thus we find today many studies of criminals, few of law-abiders; many of fear, few of courage; more on hostility than on affiliation; much on the blindness in man, little on his vision: much on his past, little on his outreaching into the future.'[1]

Personality, however, is but the threshold of a deeper prob-

[1] Gordon W. Allport, *op. cit.*, p. 18

lem still—the soul. Here we stand face to face with one of the great affirmations of religion; not of Christianity alone but, in one aspect or another, of all the faiths of man. Is there in each of us something peculiarly his own; not a self only, or even a personality but a thing of higher import still which is the real and essential person, significant and valuable in the universe itself?

Here, one may say, we must take leave of science and enter metaphysics and theology. The self is a concept foreign enough to modern psychology but the soul is so completely absent from the scientific ideas of today that merely to use the word, save as a figure of speech, marks one as almost outside the pale of ideological respectability. The soul is something that one thinks about when he faces reality by common sense in hours spent outside the study or the laboratory, but it must not be carried across their thresholds. Even Christianity today sometimes gives it little more than lip service, and the outspoken concern for its welfare nourished by our grandfathers is now found only in Roman Catholicism or Protestant Fundamentalism.

It should be recognized, however, that the evidence for a self in man, based on the organizing quality of life itself, provides religious philosophy with something tangible to which it can anchor the concept of the soul. Soul is only self writ large, only Personality invested with eternal value. We should never reach a concept of the soul by studying animals only, yet even in them there is the germ from which the soul has grown. As evolutionists, once more, we cannot deny that whatever the soul may be, it seems to have evolved from simpler biological beginnings. To draw a sharp line between organisms that presumably possess a soul (man) and those that do not (the animals) is easy to do today, but would it have been so easy in the days of Neanderthal man? Even now, on the improbable assumption that the *yeti* or 'abominable snow man' of the Himalayas should turn out to be a surviving relic of one of the ape men, perhaps *Pithecanthropus sinensis*, would we say that he possessed a soul? Would we try to make a Christian of him or would we put him in a zoo? Somewhere along the evolutionary line our theologians, if they could have been there, would have been perplexed to know just where the soul-less ape ended and man, his descendant, came to have a soul. The Church of Rome meets the problem by denying that the soul evolved, and maintains that somewhere in the past it was created, just as somewhere in the development of a human embryo it comes into being. This is a dogma of the Church but few evolutionists, I am afraid, will accept it.

The various parts of man's psychical life—mind, self, soul, spirit—cannot be sharply separated from one another. Each of these words stresses one particular aspect of the organized, purposeful creature that man is. Nevertheless, the concept of the soul stands on a higher level than that of the other intergrating factors we have mentioned. It can hardly be discussed apart from its relation to the idea of man's higher destiny and of God. A humanist, though he would probably reject the question as irrelevant, would deny that he had a soul, certainly a soul in the common usage of that term. But in all this there is a metaphysical idea which should not be overlooked and to which our present hypothesis can make a contribution. We have been arguing for the fundamental identity of the physical and the psychical in man, deriving both of them from the regulatory, purposive character of all life. This is opposed to the ancient philosophical theory of dualism, the complete distinctiveness and separate identity of body and mind. Materialists also reject dualism but they do so by tossing out the whole idea of 'mind' and 'soul' completely and basing man's unity on the material alone. The monism for which the present argument has been advanced is one that makes the physical and the psychical coequal, two aspects of the same organized living system. The fact that when the body dies, all the psychical aspects of it seem to disappear makes us almost instinctively think of the body as primary. If both are aspects of the same unity, however, the soul (or the mind, self or personality) has as much claim to reality as the body does. It might even be the *primary* reality, a part of the creative spirit of the universe, and one may think of the body not as its precursor but as its product. Let us remember William Blake's insistence that 'Man has no Body distinct from his Soul; for that call'd Body is a portion of Soul discern'd by the five senses.'

'The soul is the highest level of that goal-seeking, integrating process that is life. It is a magnificent hypothesis, and, as any good hypothesis should do, it accounts for many facts that otherwise would be quite meaningless. Until we know far more about the unsolved problems that cluster around every living thing, we should not be too hasty in denying the possibility of its existence.'[1]

The concepts of the self, of personality and of the soul involve profound questions of metaphysics. Is consciousness a necessary attribute of the self? Is awareness of self, as of something distinct from all else, a necessary attribute? Is self 'pure being'? These and others are problems for the metaphysician and need not concern us here. All I wish to emphasize is that there is a sound biological basis for *individuality* on which higher concepts of the self can be founded.

A deeper issue also is involved—is the soul one or many? The plurality of bodies has given rise, in our Western philosophy and religion, to the concept of a plurality of souls, both here and in the hereafter. This is opposed to some of the philosophies of the East, which look on consciousness and self as singulars, the plurals of which are unknown. What seems to be plurality is illusion, a series of different aspects of the same thing. The individual soul at death loses itself in the great spiritual reservoir of the universe. The very concept of personality is a Western, not an Eastern, one. These differences are fundamental, and how they can be resolved is hard to see.

Certainly no complete description of man's nature can be made that leaves out his fervent belief that he does possess a soul. Science has gone too far, I think, when it denies that such a thing is possible. A recognition of the purposive, integrating character of life, and the still unsolved problem of what life really is and whence it came, should make the most hard-headed and objective student of biology hesitate before he insists dogmatically that the soul is nothing but an empty and discarded superstition. Said a great preacher once when asked if he had a soul: 'No, indeed! I *am* a soul but I *have* a body.'

[1] E. W. Sinnott, *op. cit.*, p. 160.

9

Creativeness

INDIVIDUALS—selves—are widely various and have been continually changing. A distinctive feature of man's long history is that he has changed; has advanced, both as an individual and in his social life, from simpler to more complicated states; has moved from what we regard as a lower to a higher, more advanced level. He has always been producing new things. In a real sense, he is *creative*. Here lie his strength and his hope for the future.

This trait in him brings to its highest development a quality present in all living beings. The concept that the essential quality in development, physiological activities and behaviour is regulation to a norm or goal might suggest that a living organism, though the seat of constant chemical and physical change, is kept at a relatively static level by automatic regulatory devices. The term 'homeostasis' suggests as much. Nothing, however, could be further from the truth. Life is a constant process of *becoming*, of *creativeness*. It is in this respect that living organisms differ most from lifeless objects. A block of stone or a planet in the sky will be essentially the same tomorrow as today and for many more tomorrows, such being the conservatism of dead matter.

Life, on the contrary, is continually changing. Much of this change is slow and due primarily to the accumulation of genetic variations during the evolutionary process. How these arise is still not clearly understood, but it is certain that even a single mutation may radically alter the pattern of growth and behaviour of an organism. The meaning of organic evolution is that change *does* occur, even though it takes a billion years to make a primate from a protozoan. The lifeless universe, on the contrary, has been the scene of little comparable novelty since the beginning. The chemical elements, inorganic compounds and even the character of suns and galaxies seem to be much the same today as when life first began. Only living things have radically altered.

The individual organism, a sort of microcosm in itself, is the seat of equally unremitting change. Though steadily maintaining its identity and basic quality, it is as steadily undergoing alteration. This is evident in the series of developmental stages through which it passes, from the fertilized egg to maturity and finally to its end. Every organism moves through a life-cycle, consisting of a definite series of consecutive steps. This changing pattern is forecast in the egg. Change is constant and inescapable. Biologists have generally paid most attention to the early stages of this life history, but in recent years the study of senescence has found a characteristic series of changes in the period from maturity to old age, both in structure and in function. Even those plants of indefinite growth which do not seem to suffer from old age are still marked with the tooth of time and at last reach their end.

In this inevitable progress through a life history what is it, we may ask, that actually changes? Embryologists and physiologists have long been recording alterations in form, structure, chemical composition and metabolic activity during the various stages or phases in the lives of plants and animals. These may be analyzed and described, but what is it that causes them? In the mature and flowering branches of the English ivy, for example, why does the whole growth habit alter, when flowering time arrives, from a scrambling vine to a self-supporting stem with a polar axis; and why are leaves now oval instead of three-lobed, as they were at first? In a growing fruit like that of a gourd, why do the various dimensions change in such a perfectly regular fashion with relation to each other so that the shape of the whole steadily alters? In human development, why is it that the proportions of head, trunk and limbs in the infant are not retained in later life but change in a constant and predictable fashion? It is hard enough to account for form itself, but to understand why it alters during development, and with such precision, is still more difficult.

What the actual correlate of such a change may be, in terms

of physical and chemical processes, is not yet clear, but if the suggestion made in these pages is a sound one, the primary change occurs in the norm or goal or pattern set up in the living system. Since the various processes in the organism are such as will tend to make growth and activity conform to this goal, an alteration in the goal itself, for genetic or developmental causes or for any other reason, will modify all else. Life's creativeness lies in the progressive alteration of its goals, though these are always centred in a living organism which maintains its persistent individuality.

Conformity to a bodily goal is reached through progressively changing steps in embryological development, the goal becoming crystallized, so to speak, in the body itself. Goals of behaviour, however, are by their very nature much less fixed and are the seat of more rapid and radical alterations than are bodily ones. Here the general goal is a continuation of the organism's specific existence, but around this, as conditions change, arise a series of minor and temporary goals which regulate behavioural details. The main direction of the road, so to speak, is set by the major goal but its particular curves and dips and rises by the minor ones. Instincts, like bodily traits, change progressively from birth to maturity. Those of the larva of an insect, or of a young bird or mammal, produce a very different sort of activity from the ones adults display. In all there is a constant change in behaviour, a continual cycle of becoming.

Among higher animals, however, where memory is stronger and the beginnings of intelligence appear, instinct may be modified by experience and the animal may learn. In domesticated forms there are innumerable instances of this, but it occurs in wild species as well. A remarkable example of it has recently been found in England. Here a bird, one of the tits, happened to discover that it could remove the paper tops of milk bottles and thus have access to the milk. Knowledge of this fortunate find slowly spread through England and even on to the Continent as birds 'learned' from their neighbours to adopt this remarkable innovation in their food habits. These modifications of instinctive behaviour by experience, however, are rarely very great and most instincts are transmitted by inheritance.

With man's far more complex brain and wider repertoire of

possible choices, his psychical life becomes enormously richer and his opportunities for creative action are far greater. His brain is no longer bound by a strict pattern of nerve connections. In him rational control of behaviour has become fully developed for the first time. The variety of goals which this allows him to seek, and thus the variety of his behaviour, is vastly greater than even in the highest animals. His memory is far more extensive and his ability to relate particulars to universals and effect to cause has made reasoning possible. Because of all this he had been able to learn much more than have the beasts and thus to profit by experience, not only his own but that of others, in the present or the past. In this way he has immensely speeded up the ponderously slow process of his own advance, now no longer due to changes in the genetic constitution of his cells, as it was in the long course of evolution, but to accumulation of knowledge and experience by which an intelligent control over his environment can be achieved. Because of all this, the possibilities that are set before him have now become almost limitless. The narrow desires and aspirations of even the most intelligent of apes fell so far short of man's, when he once passed that critical point where intelligence emerged, that he became a different sort of being. The goals which he might reach mounted like water in a flood and where they will lead him no one now can tell.

Thus suddenly—in terms of geologic time—the creative change that had always distinguished life was enormously accelerated. Man himself now became a creator and proceeded to remake the world. In his triumphant advance the development of memory has been important, for it brings the past and its wisdom into the present. Reason was also vitally essential, since it can manipulate knowledge to produce more knowledge.

But there is a third element in man's creativeness as necessary as are these, for without it they would both be limited. This is *imagination*. The wide expansion in his mental goals has not been simply an enlargement of the scope of his desires and purposes, of what he wants and tries immediately to attain. A purpose does not now always lead directly to an attempt to realize it. As we have seen, it may produce a mental act instead of a bodily one; imaginary, not actual, behaviour. Goals may remain as objects of contemplation, changing, calling up images of action in a future time or in a distant place. A wide horizon or a piece of antique furniture strongly appeals to us because it stimulates the imagination to see far more, through space and time, than the eye of the flesh can ever do. Imagination is the basis of creativeness. The beginnings of it, as of other mental traits, are doubtless to be found in animals. A horse, after a hard day's work, on turning homeward moves fast and willingly. Some dim image of ease in a comfortable stall with a measure of oats may be in his mind, though the more distant future does not interest him.

Imagination, like reason and high intelligence, is primarily a human trait. It has opened the way to progress by presenting the mind with possibilities which experience has not yet known, and thus may lead to behaviour that is novel and untried. It is often productive of great accomplishment. In our long history it has been the imaginative individual, the man 'with an idea,' who pushed the race onward. In the mind of someone, before they ever came to actual reality, were the bow, the wheel, the steam engine and the atomic bomb. The rocket ship is there already, waiting to be translated into mechanism. Imagination is also a necessary ingredient of reason itself for unless one can picture alternatives and possibilities, he cannot study the relationships of things, nor deal with mathematics, philosophy and science. For successful mastery of any rational process creative imagination is indispensable.

It is in poetry, music and the arts, however, that imagination finds its loftiest expression. Only the genius can lift his spiritual goals so high as to create things of supreme beauty, but almost everyone is able to enjoy them. Nowhere is man more god-like than in the limitless scope of his imagination, which releases him from bondage to material and finite things. It breaks a path into an entirely new world and opens before him a vista of such boundless extent that it makes of him a different being, a creator and not a creature merely. The line where man transcends the brute is more sharply drawn in his possession of imagination than at any other lovel. He is not only *Homo sapiens* but much more, *Homo prospiciens*.

All this is obvious and needs little argument. The point I

wish to make is that imagination, with all that it means for man's happiness and welfare, is part of the goal-seeking process in protoplasm, goal-seeking that results in mental acts. How these goals in the imagination arise is the biological problem of goal-setting at a lofty level. The significant fact is that imagination and creativeness belong in the company of purpose, idea, thought, self and the other aspects of man's psychical life. These are not separate things, concerning radically different parts of man, but are really one, which can be approached as a single basic quality. Though this simplifies the issues by relating them closely to each other, the underlying question of their origin still looms unsolved.

An essential aspect of all these qualities that reach their highest expression in man is that they are not static but creative. This is the great teaching of evolution—that the world is moving on, perhaps aimlessly, perhaps towards some great end, but certainly moving. Lifeless matter, even as seen in the light of the theories of modern cosmogony, has little novelty. It does much the same thing over and over again—electrons, atoms, molecules, stars and galaxies. Some differences have arisen there, but little creativeness until life came. However life arose, progressive change is always associated with it. In life, matter loses its old conservatism and begins to aggregate into novel and far more complicated molecular arrangements. Whether these actually preceded the origin of life or are the result of it is not the essential point, which is that creativeness is in some way associated with the process of life itself. Through its billion years or more on earth life has advanced from the simplest of organisms to its highest peak in man, an ascent through literally millions of new species, most of which vanished long ago. The origin of these differences, the basis of life's rich creativeness in evolution, is still not understood. Whether mere random variation is involved or some form of internally directed change is not clear. In man's history and future, and in the universe itself, is there a creative purpose or is all chance and randomness? Does man's own creativeness lead to any goal that can be discerned or is it capricious and undirected? These are questions to which we shall need to return again.

One conclusion I think is justifiable: that goals in protoplasm everywhere, in a single cell or the vastly complex system of the brain, are continually changing. Organization, the essence of life, is not static but leads to new sorts of organized systems. In this creativeness is life's chief claim to be a different thing from lifelessness. Behaviour, and thus mind, is especially fluid and creative. 'Inertia,' says Ralph Lillie, 'is primarily a physical property, a correlate of the conservation which is a recognized character of the physical as physical. In contrast, the psychical, being a factor of novelty, is the *anticonservative* property in nature.'[1] This reminds us of Whitehead's famous remark that the psychical is a part of the creative advance into novelty.

Since Darwin's day, especially, many philosophers have begun to think in terms of process, creativeness and progressive change rather than in the older and more static absolutes. Bergson, particularly, has stressed this idea. Mechanism he rejects since in it all is given at the beginning. Finalism, he points out, is equally closed, since all is given in the end that is sought. Only in the questing, exploring processes of life is there room for real creativeness. Time is of the essence. Real duration 'gnaws on things and leaves on them the marks of its tooth.'

Lloyd Morgan distinguishes between *resultants*, in which the components are simply rearranged as in ordinary chemical change, and *emergents*, where something really new appears and the whole is different from the sum of its parts. In evolution, he thinks, life is such an emergent. So is mind. Emergents cannot be predicted from a knowledge of the system in which they arise. Here is true creativeness.

For Smuts, it is holism which is the creative concept. Says he: 'This creative Holism is, of course, responsible for the whole course of Evolution, inorganic as well as organic. All the great main types of existence are therefore due to it, such as the atom, molecule, cell, organism, the great groups of plant types, the great groups of animal types, and finally the human type.'[2] Samuel Alexander reaches back to the basic idea of Space-Time

[1] Ralph S. Lillie, *General Biology and Philosophy of Organism* (Chicago: University of Chicago Press, 1945), p. 189.

[2] J. C. Smuts, *Holism and Evolution* (New York: The Macmillan Company, 1926), p. 142.

itself for the nisus or driving impulse towards creativity. From this come life and mind and even Deity itself.

Such thinkers all support what has come to be known as the philosophy of process, of change, as opposed to the static philosophy of an unchanging universe. Most of them regard not only life as creative but all nature. Whether mind, self and creativity have their roots in inanimate nature or not is arguable but certainly their most conspicuous expression is in living organisms and especially in man. Here creativeness reaches its highest level in the rich productivity of the imagination. It is as if the goal-seeking quality of life, emerging from its expression in mere bodily structure and movement, became free to explore the possibilities that lie in mind unimpeded by close involvement with matter. Here man directs his attention and the nature of his imaginings to something new, unhampered by outer compulsion. It is worth noting, perhaps, that the only aspect of mentality an electronic calculator cannot imitate is imagination. Will *any* mechanism be able to create for us, as the poet's spirit can, a vision through those 'magic casements, opening on the foam of perilous seas, in faery lands forlorn'?

Man's imagination is a unique quality, a particularly significant element in his experience. It must have its roots in life, as do all his traits; and does not its splendour and luxuriance suggest that in man's living stuff there is a *spontaneous* quality that transcends simple mechanism? Are those magnificent flights of inspiration by the poet, artist, scientist and seer, which we rightly regard as life's highest expression, simply a product of the chemistry of protoplasm? Was Shakespeare such, or Bach, or Leonardo? Was Newton, or Darwin, or Einstein? Or did these men possess within themselves a bit of the generative force to which the universe itself may owe its being? The creativeness of genius presents materialism with its most stubborn problem.

The conception that life is essentially creative is different from the rigid fixity of mechanism or the capricious unpredictabilities of vitalism. Life is not undetermined, but its determinism resides in its own nature. That we do not yet know the genesis of it is simply to say that we do not yet see how goals are established in living stuff. Laws here are operative, doubtless, but we do not understand them. The composing of a great poem or symphony partakes of the cosmos, not of chaos.

Man is a part of creation but he also has a share in the creative process. His power is limited to new arrangements and combinations of things already present and he does not share in the vast and primal creativeness of God; but in a real sense he is a co-creator in the universe. This power, the highest expression of the creativeness of all life, is surely one of the most god-like qualities he owns.

10

Freedom

THE problem of creativeness is closely involved, of course, with that of freedom and determinism. If man is physically a mechanism only, there can be in him no real creativeness and thus no freedom, for in a universe the fate of which is fixed there are no novelties at all.

The underlying questions in philosophy which man must face keep coming back in different formulations as one historical era succeeds another. The 'freedom of the will' is one such problem. Whether man has a soul is an important question, but a shackled soul is hardly any soul at all. Is man really *free*, we ask, to do what he wishes and therefore endowed with moral responsibility, or is he caught in the great snare of fate which drags him on whether he will or no? To common sense and our deep inner conviction the answer can be made with no uncertainty—we *are* free, obviously and inescapably. The successful conduct of our lives, the very pursuit of science itself, depends on a belief that we can choose our course. No hypothesis has ever been confirmed more fully than has this one by its practical results, and no belief more amply justified.

And yet this belief has been most seriously challenged. As theologians in the Christian tradition pondered the attributes of God three centuries ago they were bound to ascribe to him not only omnipotence but omniscience and complete foreknowledge of the future, which thus must already be determined. On this assumption man's course is obviously predetermined and his fate is fixed. But if he is not free to act, how can God hold him morally accountable for what he does? Over this problem men long suffered agonies of doubt and apprehension. If sinners in the hands of an angry God were predestined to sin, why should they suffer the pains of hell for something they could never help? The injustice of this was obvious but the alternative seemed to be denial that God was all-foreseeing. The horns of the dilemma were a limited God or a helpless man. The stern Puritan creed endeavoured to break a pathway through this wilderness, and its best answer is to be found in Jonathan Edwards' 'The Freedom of the Will.' This carries little comfort to ordinary men—especially those not predestined to divine grace!—nor is it acceptable to many nowadays.

In the nineteenth century another reason was invoked for denying man his freedom, not the omniscience of God this time but the uniformity and predictability of nature which science had revealed. This great conception implies that every event in the universe is rigidly determined by physical causes and cannot in the least degree be diverted from its destined course. A complete knowledge of the universe at any instant and of the laws which govern it would make possible, at least in theory, a prediction of its entire future history.

> The first morning of creation wrote
> What the last dawn of reckoning shall read.

Freedom or any degree of uncertainty, in living or lifeless nature, is incompatible with this assumption of complete determinism, which justified itself by the triumphant advance of science through the century. The problem of freedom troubled many still but the horns of the dilemma now were expressed in scientific rather than in theological terms. The deadlock remains today, but it is the custom now to look at the freedom of the will as an outdated and rather meaningless problem, one chiefly of semantics and no longer requiring serious attention from philosophers.

Such a conclusion, I believe, is far from sound. It has been proposed not because freedom no longer is important but because to reconcile it with scientific orthodoxy seems impossible. Where there is such a sharp conflict, however, between the deterministic requirements of scientific theory and an assumption of freedom based on almost universal human conviction, an assumption which underlies most of our philosophy and morals, every effort should be made to reconcile the two. The problem of freedom is by no means academic only. Many of our ills today are the result of a growing acceptance of the idea that we are puppets, pushed around by fate and circumstance and no longer masters of our own destiny. The moral and philosophical implications of such a belief are momentous.

Aside from the general question of physical determinism and the uniformity of nature, there are two more immediate sources—psychology and genetics—from which the conclusion has commonly been drawn that man is not free. They often oppose each other but each has many supporters.

Psychology, like science in general, is so committed to determinism that although it often speaks of choice and purpose it uses these terms largely for convenience and sidesteps the basic problem. Sherrington comments that 'from the human standpoint, the important thing is less that man's will should be free than that man should think that it is free.' But until the science of the mind can develop a foundation of theory that can satisfy its own data and at the same time bring them into harmony with the universal conviction of mankind and thus reinforce his sense of moral responsibility, it will fail in its chief service Somewhere, somehow, the barriers between theory and experience must be breached. Professor Allport speaks wisely about psychology's attitude towards freedom, though perhaps for a minority of his colleagues.

> No other issue [he says] causes such consternation for the scientific psychologist. One may look through a hundred successive American books in psychology and find no mention of 'will' or 'freedom.' It is customary for the psychologist, as for other scientists, to proceed within the framework of strict determinism, and to build barriers between himself and common sense lest common sense infect psychology with its belief in freedom. For the same reason barricades are erected against theology. But to our discomfort recent events have raised the issue all over again. Existentialism insists on freedom; much of the psychotherapy now in vogue presupposes it; psychology's new concern with values is at bottom a concern with choices, and therefore revives the problem of freedom. Up

> to now the tug of war between free will and determinism has been marked by naïveté. Just as we have learned with some success to transcend the monolithic oppositions between mind and body, nature and nurture, we should strive for better perspective in our view of freedom and determinism.[1]

At present a substantial body of psychological opinion looks on man as an 'empty' organism, a *tabula rasa* on which his environment writes everything and he, nothing. On an earlier page this position was presented briefly. To quote Professor B. F. Skinner again:

> Man, we once believed, was free to express himself in art, music and literature, to inquire into nature, to seek salvation in his own way. He could initiate action and make spontaneous and capricious changes of course. Under the most extreme duress some sort of choice remained to him. He could resist any effort to control him, though it might cost him his life. But science insists that action is initiated by forces impinging upon the indvidual, and that caprice is only another name for behaviour for which we have not yet found a cause.[2]

For Professor Skinner, freedom is mere caprice. Unless it turns out to be more than this, however, we may well agree that it is meaningless. No one will go to the barricades to defend mere randomness and aimless whimsey. We can agree that *something* determines what we do. The great question, standing at the very foundation of our inquiry into man's nature, is whether this factor comes entirely from outside man himself, determining everything he is and does, or whether there is something within him that shares in the directiveness. No one can doubt, of course, that much of our behaviour is the result of influences in our environment—upbringing, educa-

[1] Gordon W. Allport, *Becoming* (New Haven: Yale University Press, 1955), p. 82-83.

[2] B. F. Skinner, 'Freedom and the Control of Men,' *American Scholar*, 25:52-53. 1955-56.

tion, culture pattern and conditioning. The son of an Episcopalian is likely to be an Episcopalian and of a Democrat, a Democrat. One brought up in a family of wealth will acquire different economic views from those of a labourer's son. The pattern of culture south of Mason and Dixon's line imparts a different attitude towards race relations than one common in the north.

A belief in the supreme importance of environment is adopted by much of sociology today. We are told that the way to improve mankind is by giving us better conditions under which to live—better food, housing, medical care and other things that minister to safety and well-being. Many students of ethics maintain that morals are imposed on us by our culture and are without any inner authority. Criminals are not really culpable but simply the unfortunate victims of a poor environment.

Certainly much hope for man lies in the proper manipulation of his environment, for he is very sensitive to it and is highly educable. So to condition him that he will co-operate with his fellows unselfishly and value qualities that will make him healthy, wise and productive offers to the practitioners of applied psychology a challenge to remake the world. This has a tremendous appeal, for it is essentially the same challenge that has inspired teachers and reformers and missionaries and evangelists from the beginning. What is new in it, however, is the assumption that these external agencies are all-important and that the individual himself has no control over his own behaviour. The freedom he seems to possess is only illusion.

But in their conviction that factors outside the organism are of paramount importance in determining its behaviour, psychologists have too much neglected the organism itself. It certainly is by no means the empty, neutral thing that many assume it to be. For over half a century the science of genetics has accumulated a vast amount of evidence that what an organism is and does depends to a great extent on its inborn hereditary constitution. Two individuals may react very differently to the same environmental factor. Less is known about the genetics of man than of many lower organisms because of the difficulty of getting extensive data, but it is certain that the general principles of heredity apply to him as fully as to other living things and not only to his physical traits but to his mental ones. This does not mean that inheritance is everything, for environment obviously exerts a profound influence. It is well recognized that what a gene controls is not a specific trait but a *specific way of reacting to the environment*. Both heredity and environment are major influences on an organism, and neither should be emphasized to the exclusion of the other. Let us not forget, too, that protoplasm is not a nonspecific system. It has a physical and chemical constitution that is remarkably uniform throughout the organic world. We should expect it to have certain qualities and predispositions of its own which will resist the influence of either environmental or genetic factors and such, I believe, is actually the case.

Some geneticists are as dogmatic as some psychologists. As the latter would try to persuade us that man is a puppet and under the complete control of his environment, so the former, especially when they are considering the human species, stress the paramount importance of inheritance. Professor C. D. Darlington [1] in a recent book tries to convince his readers that a person's hereditary constitution determines most of what he is and does. There is indeed some very striking evidence of the often subtle role heredity must play, notably in cases of identical twins who have been reared apart but in whose lives events are remarkably parallel. Most geneticists, however, would not draw such sweeping conclusions as does Professor Darlington. This book will serve as an excellent counterpoise to one by Professor Skinner [2] for both are extreme statements of their particular points of view.

Either of these theories, however, or any division of man's fate between them would mean that freedom is drastically abridged or wholly lacking. One can be as much the slave of his genes as of his conditioning. Indeed, genetics has more often been cited in support of a deterministic philosophy than has psychology. Certainly neither of these sciences, at least in their

[1] C. D. Darlington, *The Facts of Life* (London: George Allen & Unwin Ltd., 1952).

[2] B. F. Skinner, *Science and Human Behavior* (New York: The Macmillan Company, 1953).

more thorough-going applications, seems likely to provide a theory of behaviour that will be scientifically acceptable and at the same time offer a satisfying interpretation for man's conviction that what determines his behaviour, at least in part, is something within himself over which he, as a person, has control. One can hardly expect this ancient question to be settled now by any new insight, but it certainly should not be neglected or suffer from a conspiracy of silence. It is here, perhaps, that the suggestion offered in these pages may prove useful.

Causation is an abstruse subject upon which we need not enter here. It is clear, however, that our acts are not random happenings but follow some sort of orderliness, some kind of law, else chaos would be supreme. If the laws of chemistry and physics with which we are familiar and which are the basis of mechanistic philosophy do not explain what we do, then some new principle must be invoked. We should not be afraid to explore widely in the field of unconfirmed and unorthodox ideas for new principles of this sort, even if what is found upsets some cherished preconceptions. Biological organization may be considered as such a new principle. What, we may ask, can it do for an interpretation of the problem of freedom?

Its contribution is a simple one but not, I think, without value. This centres in the concept of the self, of what each of us actually is. The self-regulation of protoplasm, its basic quality, means that all its activities tend to realize goals—in man, goals of many kinds in the same individual. The important fact is that the self is *the sum of all the goals of the individual*, all its purposes, desires and aspirations, the total of its organizing relations. We *are* what we *want*. Wanting—goal-seeking—is at the centre of biology. The self is the essence of man's aspirations, of the things he seeks. It is not a static, nebulous, ghostly thing but a search and a desiring; a pattern in living stuff which seeks to be fulfilled.

If this be once admitted, it seems to me that the problem of freedom has been moved a long step onward towards interpretation. Since the behaviour of an organism results from the tendency of its living system to conform to goals or purposes that are set up within it, and since the self is the sum of these purposes, the self is obviously a part of the means by which purpose is translated into deed. The purpose and the purposer are one, and both are aspects of the regulatory power of living stuff which moves it to the consummation of the purpose. If this is so, a purpose in the mind is not something alien to the deed, a different category of things; part of a system about which we can argue as to whether it is 'free' to do the deed or not. In a sense it *is* the deed for it is our experience of the protoplasmic goal which the deed will realize. There is no *compulsion* here to act, for the very act is an expression of the self. Obviously, we do what we will to do. Among the various choices open to us, we make the one which we want most to make. Freedom is the coincidence between ourselves and our acts.

Of course not every want can be satisfied or every goal realized. There are forms of outer compulsion which bind us and prevent our doing many things we want to do. Lack of physical strength, the determination of others who are more powerful than we, absence of opportunity and many more. These are life's familiar frustrations but are not limitations to our fundamental freedom to choose our course. Against these obstacles our purpose holds. Like the equifinality shown by some embryos where, if one course of development is blocked, the organism reaches the same end by quite a different route, so a man whose motives are strong enough will try one course after another to overcome these obstacles and reach his goal. His will is free but cannot always bring to pass the thing it wills.

There are other scientific arguments for the conclusion that our choices are free ones. Chief among these is one from evolution. Why should there exist this whole process of apparently free, conscious choice among presumable alternatives unless it was of some advantage for survival? The sense of obligation, of what we *ought* to do, implies not only a moral imperative but the ability to choose between right and wrong. Feelings of indecision, anxiety, regret, remorse and others, which so often make us miserable and are such a conspicuous part of what we think and feel, should be something more than useless by-products of protoplasmic processes. They must be part, one would think, of the evolutionary mechanism.

This argument applies with particular cogency to the old problem of pain. Pain is a warning that something is wrong with our bodily organization—injury, infection or malfunctioning of some part of us. Without this warning we should not know the peril until too late to seek a remedy. To be effective, the warning must cause us to change our course of action, thus implying that we are free to do so. Surely some automatic and unconscious painless remedy could have been developed for these ills during the course of evolution, as has happened so often in the physiological processes of the body—the production of antibodies, the mobilization of phagocytes and many other activities of the *vis medicatrix naturae*. If we are really automata without freedom of choice it is monstrous indeed that we should have to suffer not only physical pains but the more subtle ones of anxiety and remorse. The problem of pain and evil is an ancient one and difficult, but we must conclude that it would be indeed a poorly planned universe where a helpless machine must be condemned to suffer.

What determines our choice involves the problem here discussed so often—how, in an organism or within the brain, are protoplasmic goals for action set up? Until the laws which govern these are understood the concept of freedom will still remain incomplete. Here is the nub of many problems. There must be *some* sort of physical correlate in the process of choosing a course of action, and that correlate, I believe, is the protoplasmic pattern to the realization of which its activity conforms. That something more than the classical laws of matter and energy may here be involved is suggested by the remarkable developments in the physical sciences during the past half century. This is an active field of debate by philosophers and theoretical physicists today but so full of technical complexities that a layman has difficulty in understanding all the points involved and should not draw from it conclusions which are not supported by those who are able to speak here with authority. Many of these men, however, have become convinced that the doctrine of simple mechanical determinism has passed into limbo. The radically new ideas which produced the great revolution in physics encouraged open-mindedness on the part of physicists. This open-mindedness caused them to entertain ideas about the universe and its history and inhabitants which would have been cast aside a century earlier as violating the basic conclusions that science had been so long in establishing. To such men, physical indeterminism is not as shocking to contemplate as it had been to their predecessors.

One of the early indications of this change in attitude had to do with natural law itself, the very citadel of science. Many kinds of law—some say all—are now regarded not as rigid and inviolable necessities in the universe but rather as the large-scale result of the random behaviour of a vast multitude of tiny physical units. At temperatures above absolute zero, for example, material particles in a gaseous or liquid state are in constant motion and are continually colliding and rebounding. This leads to their diffusion. The course of an individual particle seems completely random. Ultimately, as the second law of thermodynamics states, there will be a very high degree of probability that, as the result of this random pushing and scrambling, matter will be evenly distributed in any given space. Out of this random behaviour certain predictable uniformities appear which can be expressed as laws, but simply as *statistical* laws. A bit of gas, liberated in the corner of an empty container, will soon spread evenly through the available space because of the continual collisions between its particles and will exert a predictable pressure on the walls. This does not *necessarily* happen. It is possible, though most improbable, that the gas might remain in a corner of the container. Actually, this chance is so slight that upon this random particle behaviour we can build a law such as Boyle's law and be confident that it will hold. The practical difference between a statistical law and an absolute one is negligible, but the theoretical difference is great, since the former admits a degree of indeterminism—tiny in amount, to be sure—into the workings of the universe.

Studies in the behaviour of the smallest individual particles of matter support this conclusion. The principle associated with the name of Heisenberg states that it is impossible to determine both the location and the momentum of an electron at the same time, thus indicating that firm prediction about the behaviour of such a particle cannot be made. Closer to our problem, perhaps, are observations made on a stream of electrons in the beam produced by an electronic accelerator. The direction of a large beam, composed of countless minute particles, can be accurately predicted. As the beam is made smaller this accuracy decreases, and finally the path of a single electron seems to be quite unpredictable.

This problem of determinism in physical law is a vast one and cannot be discussed here in detail. Its bearing on the idea of freedom in man's behaviour is important, for the suggestion comes to mind that this minute degree of freedom present in matter might be amplified, so to speak, into freedom at a much higher level. There is peril here, of course, of translating into human freedom the indeterminism of randomness, for the latter tends towards disorder. What we seek is the basis of the freedom of an organized, goal-seeking, purposeful system. The fact that material particles seem not to be rigidly determined offers hope that the behaviour of the ultimate particles of protoplasm may escape bondage and be free to manifest the operation of other laws which deal not with lifeless or uncomplicated matter but with those highly organized systems that we call living organisms. Such laws, still undiscovered, may open the way to levels of truth above those with which we are familiar in simpler forms of matter and energy, and lead to an understanding of those psychical qualities in man that are such a vital part of his nature.

The theory of determinism itself is by no means as rigid as it is often assumed to be. The orderliness of the universe, on which science depends and which it continually seeks to explore, rests on one great assumption, amply supported by experience—that the same conditions are always followed by the same results. This is experimental determinism, the only kind with which science is concerned. What happens depends on the conditions, and when the conditions are different, the results will also be different. New conditions often have new and unpredictable consequences. This is a modification of the idea of a completely fixed and mechanically determined universe. Among the varying conditions are the goals of an organism, the expressions of complex protoplasmic configurations. These are almost infinitely various and it is therefore almost impossible to predict what an individual will do. He will follow his purposes, the directives of his mind. If purpose determines action, is not this what we mean by freedom? How goals change, whether by environmental factors or genic ones or others, is the final problem. As new conditions appear in biological systems there may be evident new qualities which could not be known before. This opens a way for freedom—not complete freedom, perhaps, nor even much, but enough to loose us from the shackles of a strict determinism.

The difficulty frequently experienced of providing a scientific justification for the concept of freedom is doubtless in part responsible for the feeling in many minds today that freedom is not the indispensable thing our fathers thought it was; less to be desired than security, a pleasant life and intelligent supervision of our doings. Why should we fight for freedom if actually we are puppets anyway? But despite the defeatist attitude of a minority and its inculcation by totalitarian theory, most men still eagerly want freedom. They believe that they *can* break their chains and choose their course and stand on their own feet, and that no clap-trap about determinism, psychological or genetic, can make them slaves to matter or energy despite all theories about the unbreakable sequence of events. Their inspiration comes from men like Milton and Jefferson and Lincoln, not from Marx. Man is never so like a god as when he gives up everything else for freedom. Life would be a ghastly travesty if the freedom for which through the ages men have lived and died proves, when at last they win it, to be mere illusion. What profit is there if they overcome external tyranny only to find this triumph meaningless since they are bound in the unyielding and impersonal web of matter and energy, and freedom turns out to be a mockery?

> For what avail the plough or sail,
> Or land or life, if freedom fail?

The deep question about freedom is, *what* is free? A freedom that was mere capriciousness would have no value. If something beyond genetic determinism and environmental conditioning controls our acts, what is this something? I believe it is the inborn, native tendencies of life; the creativity that is inherent in every organism and most especially in man; the inner protoplasmic directiveness, moving towards goals of many kinds and at many levels. To seek an answer to this question we must therefore consider in more detail the highest aspect of man's goal-seeking.

11

Spirit

PROTOPLASMIC goal-seeking, as manifest in the regulation of plant and animal growth and activity, is an essential concept for biology. Since it affects the purposiveness of behaviour, it is the basis of mind and thus an essential concept for psychology, as well. Beyond this it can make, I think, some contribution to a problem deeper still, the nature of man's *spirit*. This aspect of him reaches over into the realm of metaphysics and religion, and here our speculations move into questions which, though rooted in biology, finally pass beyond the competence of science.

That such a thing as the human spirit exists at all will be denied by many. It seems to them an idea so flavoured with mysticism, so reminiscent of those ancient, unenlightened days before man's growing reason had freed him from superstition, that it no longer is worthy of serious attention. To such objectors we may reply that his psychical life has various aspects, and that one of them—by many thought of as the most significant of all—needs a particular name by which it can be distinguished. In the sense in which it will here be used, the term 'spirit' need have no mystical overtones and certainly no taint of superstition. It is simply one special aspect of man as a living being.

Goal-seeking in a primitive mind translates itself directly to behaviour. What a savage wants to do he *does* at once if it is possible; but since far more goals exist within his mind than can be gained, or even pursued, at the same time, most of them stay latent there as things desired, to be gained at some time in the future by appropriate acts of behaviour when opportunity offers. He likes eating meat, let us say, and enjoys the pleasures of the chase or the grimmer ones of combat. The company of his mate delights him, and the warmth of the family fire. These are all goals which from time to time he tries to reach, but when he is not doing so they linger in the background of his mind as things he would like to do. In intellects more refined, such goals become translated into clearer images of things that might be, a body of desires, hopes, interests and ideals. They interfuse most of man's mental life. The wish in a real sense thus is father to the thought. Our minds are coloured by the presence there of such goals, unrealized, often unconscious, but powerfully affecting what we do. As the magnetic pole pulls the needle of the compass towards it, so they direct the course of our thoughts and actions. They form the basis of our *values*.

The experience of these desires is rarely a neutral and passionless feeling, a cool and intellectual affair. Sometimes, indeed, it seems to be such, as is his who comes to appreciate a masterpiece of art or literature by dissection and analysis, or who embraces a religious creed from purely rational conviction. Far more commonly, however, these cravings transcend mere rational justification. They are experienced with a warmth, intensity and vividness that are the source of those strong feelings of attraction and aversion which are such vital elements in human life. Goal-seeking and purposiveness at this level are transmuted into emotions—love, hate, fear, aspiration, anxiety, delight in beauty, reverence and many more. They are what chiefly distinguish man from a machine and give to his existence the zest and fire and richness that lift it so far above mere rationality.

The emotions of primitive man were doubtless those that had proved advantageous for his survival. Darwin described many instances of this. Despite the bloody evolutionary road that man has travelled, however, a remarkable thing about these emotional attitudes of his is that they include not only those appropriate for a species that has fought its way up, but that so often they are directed to qualities and objectives of a different sort, ones that the universal judgment of the race regards as higher than mere brutish goals. As man grows out from barbarism there rise in him, wherever he is or to whatever race he may belong, cravings for beauty and rightness, love and truth, reverence for something greater than himself that dwells in nature. His interpretation of these qualities may differ widely with time and culture and geography. For a while he may abandon them in part and slip back towards savagery again, but throughout his history man always has returned to these as to his highest values. At his best, something within him seems *naturally* to seek such goals. This conclusion may seem overly optimistic for a time like ours, but in the perspective of history I believe it can be justified. Like Christian in his pilgrimage through Bunyan's allegory, he always lifts his eyes to something higher.

Emotions are subjective; sensed within, and expressed outwardly only by their physiological accompaniments—the faster pulse, the quickened breath, the flushing of the cheek. Nothing like them, we may safely say, happens to a machine, even the most complex. Their beginnings can be inferred among the higher animals, but emotions are far more varied in man's life and mark him off from all the brute creation. He still has animal appetites and passions, to be sure, but these higher cravings are peculiarly his own. Hunger for food he knows, but also hunger for the very different nourishment of beauty. Sexual passion is his, as in the brutes, but he can transform it to the noblest of the emotions, love. Pouring up to consciousness from his unconscious mind, that deep source of purpose and desire within him, these loftier emotions mark a level that is higher than physical goal-seeking and quite different from intelligence and reason. 'These deep-seated inborn urgencies and desires, arising spontaneously in the mind but subject to a wide measure of direction, often dragging man down to the level of the beasts but coming to flower as the highest expressions of what he is and what he might become, one may rightly call, I think, the human *spirit*.'[1]

Let us define our terms here somewhat more precisely. What *is* man's spirit, really, we may ask? How does it differ from his mind? his self? his soul? More than all, is it related to any spiritual thing outside, perhaps to a Divine Spirit that dwells in nature?

These are questions impossible to answer satisfactorily. Since

[1] E. W. Sinnott, *op. cit.*, p. 127.

what we have called the spirit is an aspect of the living human being, however, the sciences of life and mind should not pretend a lack of interest in it. Spirit has been such a vital element in the life of man from the days of witch doctors until our own that whatever the biologist can say about it should be welcome. At best, this never can be very much, but the foundation that the man of science here can lay may serve for an edifice that will shelter not only himself but the philosopher and the man of faith, as well.

The various expressions of man's psychical life cannot be very sharply distinguished. *Mind* we may assume to cover all the psychical aspects of the human organism. The *self* is his particular individual being as he experiences it. The *soul* is the essence of the self, the precious part of man, looked on by many as his immortal portion. *Intelligence* is that aspect of the mind which sees the relationships of things and draws from them logical conclusions. *Spirit*, as the word here is used, denotes an aspect of the psychical whole that is different from any of these, though not sharply marked off from them. It is like instinct but at a higher level. It includes the stream of deeper feelings and emotions that stir within us, but it is more even than these. It is not mere stirring, mere emotion, but is directed *towards* some object—admiration for a work of art or a piece of music that draws us on to see it or to hear it played; delight in a poem that sets us to reading it; love of a person that makes us want to be with him or to think of him; hatred or fear of something, that fills us with a desire to crush it or avoid it; reverence for whatever is awesome and sublime that brings us to our knees in worship. These feelings may occur at many levels of intensity, from peaceful contemplation of a sunset to the fiery passion of an evangelist or the mystic's ecstasy. In all these cases there is feeling, a stirring of the heart; but unless this feeling is directive, unless it leads us towards that object by which the feeling is aroused, unless it *aspires* to something, it is not a manifestation, I believe, of what is rightly to be called the spirit.

This quality brings the spirit down, you see, to that same biological fact of goal-seeking and directiveness that we have been discussing. This is a common denominator that binds together all manifestations of life and makes of an organism, even a man with his almost infinite diversity, a single whole. Through the blindly questing embryo that builds random matter into a patterned body, through the marvellous regulations of behaviour that one sees in instinct, and through the orderly purposiveness of rational thought, this goal-seeking runs like a golden thread. Why should we not recognize the same fact in the more mysterious realm of deeper, unreasoned aspiration? Since man here, too, is seeking to reach goals, must not these goals be foreshadowed in some way within the organization of that stuff which is the basis of his very life? Whatever spirit is; whatever it may mean for an understanding of man's nature and the problems of philosophy and of religion; whatever it may finally portend for his hope and destiny, it has a relationship with matter through its birth in the regulatory processes of protoplasm.

If this conception of the spirit as rooted in protoplasmic regulation is a true one, it can do something to bring this supreme but nebulous part of man down from the cloudy regions of metaphysics and theology and at least put a biological foundation under it. Bergson rightly says, I think, that the life of the body is on the road to the life of the spirit, and that we have erred in the past by trying to isolate spirit from the rest of life, thus making it unreal and ineffective.

This conception of the spirit, however, ingenious though one may admit it is, will certainly not pass unchallenged. Philosophers of both materialism and of religion may be expected to object vigorously to it, though on quite different grounds.

The evolutionist will hasten to point out that the emotions—for in the end, he says, it all comes down to that—have had their origin in the slow course of man's ascent. They have been fashioned by competition and natural selection, just as have all other human traits. Did not Darwin himself write a book on the expression of emotions in animals which showed their gradual development among the vertebrates and their usefulness for survival? Love, for example, is either an extension of the reproductive urge which brings the sexes together and without which life would not continue, or of the maternal instinct so necessary in the nourishment and protection of the young until they can care for themselves. Even in human societies, family solidarity and the ties that bind individuals into a common group are of great value for survival. The warm emotion of love has a necessary and reasonable place in the life of man, and to explain it no mysterious 'goal' is required. Moral qualities in general are of such obvious social value that they would have been developed for this reason alone. And, at the other extreme, rage, hate and fear are accompanied by physiological reactions (in glandular activity and in nervous and muscular changes) which increase the individual's combative potential or aid him to escape from danger. Their evolutionary usefulness is obvious.

Other emotions receive similar explanations. Courage, curiosity, sympathy, anxiety, shame and others may all be understood as serviceable traits. Even where their immediate value for survival is not obvious, if such qualities are present it must be assumed that at some time or other, in the complex processes of competition and adaptation, they have proved their usefulness.

The physiologist, too, can present abundant evidence that for several of the emotions, at least, a definite biochemical basis has been established. In anger, adrenalin is poured into the bloodstream and stimulates muscular activity. Prolactin, the 'mother-love hormone,' is clearly related to the display of maternal instincts. Drugs of various sorts produce important emotional effects and may even lead to ecstasies and delights resembling those experienced by mystics. For all such, as presumably for every activity of the mind, there is a chemical cause.

Emotions are the particular province of the psychologist, and he joins his scientific brethren in assuring us that there is a definite physical basis for them. They are related to activities in the thalamus. As in all mental states they are derived from the complex interplay of stimulus and response. Often they are the outcome of conditioning, as when a shock of fear is produced by an air-raid signal even after war is long since past. Much is now known of the causes of the emotions, and psychology denies that there is anything which might be called mystical or supernatural about them.

In answer to these objections it may be replied that in the suggestions here proposed as to the origin of spiritual qualities nothing mystical or supernatural is assumed, in the sense of facts outside the realm of law. Unexplained they may be but surely not to be thought of as having no possible explanation. The germ of the emotions, and thus of qualities that have here been termed spiritual, is certainly present among animals. In the complex adaptiveness of life they doubtless have often been of value and preserved by the action of natural selection. There are many instances, however, where their usefulness has certainly *not* been proved. Darwin himself sites cases of helpfulness between members of different species, which would hardly provide material for natural selection. The many instances where animals seem to be amused or to experience enjoyment could hardly be so explained unless a happy animal is more likely to survive! The desire for beauty seems especially hard to attribute to the influence of natural selection. In man's history, certainly, it must often have been a hindrance, not a help. Many high civilizations that paid homage to it and diverted some of their energies thereto have been toppled down by rude barbarians who kept their weapons bright. Dedication to a moral code that stresses altruism and self-sacrifice may actually be a handicap to survival in a rough and selfish world. At all events, knowledge of evolution is still so incomplete that we are by no means justified in assuming that *every* trait that exists must do so by virtue of some usefulness that it possesses. We certainly would not be willing to accept this explanation in the field of social evolution, or to agree that 'whatever is, is right.'

The problem of the physiological and psychological bases of emotions, and thus of spiritual qualities, is simply a part of the ancient question of the body-mind relationship. At least it may be argued that if the material environment, inside and outside the human organism, is such that invariably there are produced in men those attributes that we have here called spiritual, my main argument—that these qualities are inherent in man—is essentially conceded. Although no confident philosophy of mechanism will admit that the physico-chemical system that man is can ever have qualities that deserve the name of spiritual, the human *spirit* remains the most embarrassing fact that this philosophy must face, for it implies either a spiritual element in man or the emergence from matter and energy of high qualities that materialism has no reason to expect to find there.

The objections to the hypothesis from the other side, the advocates of a religious philosophy, are no less vehement. To a man of faith the conception of his spirit as simply the sum of his emotions and desires seems so inadequate as to be almost meaningless for an interpretation of religion. Spirit, he says, is something that pervades the entire universe. Men are born of it, are filled with it and walk after it. God himself is spirit. What connection can possibly exist between these august facts and any processes, however curious and interesting they may be, that can be found in such a material system as protoplasm is? The two belong to different universes of discourse. 'Those that are born of the flesh are flesh, and those that are born of the spirit are spirit.' The teaching of the gospel is that we must be lifted out of our physical, mundane, biological existence and be born again into a new and *spiritual* world. It is neither good science nor good religion to talk about a biological basis for the spirit. Emotions are important and when rightly used can be of value in man's religious life; but Spirit, the word with a capital letter, the key to man's salvation and the seal of his true nature as a child of God, the great creative Presence that in the darkness of the First Day moved upon the face of the waters—this surely is so far above the level of biology or psychology or any other science formulated by the wit of man that it is preposterous to think of them together. These objections I shall try to answer on a later page.

And at this point the argument temporarily may rest. The position here defended is briefly this: that neither protoplasm itself nor man, its highest expression, are neutral systems in which whatever happens is the result only of outer forces, but that specific inner ones are also effective; that the character and organization of living stuff is such that goals are set up within it which are characteristic and specific at all levels, though subject in their particular quality and effect to the influence of the environment; that the progressive advance in these goals during evolution, though in large measure certainly the result of natural selection, has been due in part to specific tendencies in protoplasm itself; and, most important of all, that the direction in which these goals have advanced is such that in man they tend to create the high ideals of beauty, right, truth and the Divine which in a real sense may be called spiritual. Spirit in man is a result of the autonomous, creative quality that is his. Like the wind, it moves where it lists; but whence it comes, and whither it goes, we do not know.

If these ideas are correct, conclusions of some moment can be drawn from them both for philosophy and for religion.

First, the goals set up within our living stuff, from its lowest level to its highest, determine the course of our desires. When not translated into action, these goals, as we have seen, become latent in the mind as wishes, aspirations and ideals. Man is not neutral about the environment in which he lives. Some things he likes and some things he does not. He is continually making *value judgments* about them. Values are things about which one hears much today. These are of many sorts and at different levels, from purely physical preferences and desires to those we may call spiritual. Values are goals. The most distinctive traits a person has are not so much what he is, in body and mind, as the things and qualities he values. These, in a sense, *are* what he is. His physical life, his behaviour, his philosophy and his religion will finally depend on what it is he *wants*. This is the basis of his self. In any attempt to learn what man's real nature is, a study of his values is therefore of paramount importance. If these are imposed upon him entirely from without, he may truly be said to have no real character of his own, for what he is bears only the stamp of his environment. If they arise within himself, however (modified though they may be by many factors), they are a prolific and dependable source of information about him.

The fact that everywhere, as man has advanced from barbarism, there grew in him the high spiritual values we have been discussing surely tells us something that it is significant as to what he is. Men differ greatly in their values. In some of them these never reach a high livel, or if they do, they are crowded out by lower ones. For most people, however, spiritual goals, though never completely realized, remain as the supreme values of their lives. That man loves beauty, seeks righteousness, pursues truth and reverences the Divine, and does so not primarily from outer compulsion or inner necessity but because these are things he earnestly *desires*, should put us in good heart about his future. A creature with ideals like these must have rudiments within him of something that is truly god-like.

It is just here that one who seeks to build a bridge between the physical life of man and that higher existence of the spirit which is religion's theme can find a place to put it. Such a bridge exists, I think, in protoplasm, for this is the point where the material complexities of bodily life are so co-ordinated that goals arise in it—by what means we know not—which, like surveyors in a wilderness, mark out the course that it will follow. This started with the development of the body, passed on through the growth of the mind and the triumphant rise of reason, and finally pushed out into the mysterious territories of the spirit. It is in this goal-seeking character of protoplasm that matter and spirit meet. Whether spirit is born here from the complexities of matter or whether spirit at this critical point becomes associated with it and directs the course that matter takes is an important question for philosophy.

No one has yet an answer to this problem. The point I wish to make is that the case for spirit as a specific reality is stronger than has generally been admitted. It need not rest alone on inner conviction, faith and the authority of revelation, important though these may be, but can look to the basic fact of biology—goal-seeking—for evidence of something that may truly be called spiritual. This position, to be sure, must assume that life and at least the rudiments of spirit are co-extensive; that life, mind and spirit essentially are one. To understand the nature of life we shall never succeed, I think, by seeking to interpret it solely through the laws of physics and chemistry which we now know, but we shall need to discover new laws, perhaps a new *kind* of law, for life. When we find these, we shall not only solve the deepest problem in biology and the ancient enigma of the relation between mind and body but shall also dimly begin to understand how spirit, mysterious as it now seems to be, can come to dwell in the flesh.

Important though emotion is in the spirit, the *highest* manifestation of spirit is far more than goal-seeking. It is a moving experience of desires and aspirations carrying an inner certainty of their own high character and with power at times to lift us out of ourselves. Men of all qualities and kinds, though never many of them, have had high mystical experiences and report upon them with surprising unanimity. As William James has said, 'In Hinduism, in Neoplatonism, in Sufism, in Christian Mysticism, in Whitmanism, we find the same recurring note, so that there is about mystical utterances an eternal unanimity which ought to make a critic stop and think, and which brings it about that the mystical classics have, as has been said, neither birthday nor native land.'[1]

A sense of exaltation, a longing for an intimate communion with a greater Spirit felt to be near and a certainty of conviction that the experience, brief though it usually is, has given a true picture of reality, is felt by all of them. Such high ecstasies are the lot of only a few, but lesser ones are the common experience of the devotees of all religions. They are the loftiest expressions of spiritual goal-seeking. The foundation of religion is the conviction with which such states report the existence of an unseen world with which it is possible for man to hold communion. Whether these spiritual insights tell the truth or are mere illusions has long been debated, but a substantial number of philosophers and even a psychologist or two like William James believe that they lead us at least into a realm of reality not accessible by reason alone. This means, in the terms we have been using, that these high goals that are set up within us come somehow into contact with the same sort of spiritual reality in the universe outside. Such is the great leap of intuition that religion has always made, an adventurous but confident essay of faith that it *is* possible to link our physical world with a far richer one of spirit inaccessible to man's senses. That such a leap may not be completely blind is suggested by the hypothesis presented here. Spirit is but the highest expression of *life*, and life is still as unexplained as spirit is. The final problems of biology and of religion may well turn out to be the

[1] William James, *The Varieties of Religious Experience* (New York: Longmans, Green & Company, 1911), p. 419.

same. The vivid inner experience of life at its highest in these aspirations and spiritual insights born in the processes of protoplasm throw light on the operation of that remarkable organized living system we call man.

Man's spirit is first of all a questing and aspiring thing, seeking in the world outside for something to satisfy its inner longing. That there exists a means by which this satisfaction can be gained—a reservoir of spirit on which a man may draw—is suggested by the very longing itself. Nature is not frustrating. For every goal that draws us on, the thing desired exists. That we are hungry or cold implies that food and warmth are to be had. Sometimes it may be that we cannot reach them, but they are there. All lower goals set up in protoplasm can be attained. It would be strange indeed if men were continually tantalized by urgent desires within them which could never be fulfilled. Does not the existence in our hearts of longings for spiritual satisfaction imply that means for satisfying these indeed exist? Must we believe that only here is an explorer deluded by a mirage and that there is nothing that can quench his spirit's thirst? Thus to frustrate men is never nature's way.

If all this is so, religion's ancient insistence can be justified—that insights of the spirit give us a valid picture of truth through a knowledge of it quite different from that which intellect provides. Spirit is the part of man that seems to be in immediate contact with reality and that can feel directly what its very substance is. In many things, of course, rational intelligence is the only safe highway to the truth. But where this no longer can be followed, as in the deepest questions that men face, the other road of spiritual insight, says religion, also provides access to it, and in the end is the most reliable of all. If man's spirit actually *is* in contact with what William James calls 'a more of the same kind outside,' if man's spiritual goals are not empty dreams but directed towards things that actually exist, then religion's stout assertion has the support of something besides faith alone.

To the materialist and the positivist this conclusion is the veriest moonshine. For them, report of sense and reason are the only verities, and spirit, with whatever it may claim to reveal, is all illusion. These inner feelings and desires, they say, of which we here have made so much, are notoriously subject to outer influences—to oxygen supply, the liver's health, blood chemistry, drugs and many other things. To build a philosophy on evidence from such a source as this is fantastic foolishness. The words of Thomas Henry Huxley still express the opinion of many when he remarks that 'anyone who is acquainted with the history of science will admit that its progress, in all ages, meant, and now more than ever means, the extension of the province of what we call matter and causation, and the concomitant gradual banishment from all regions of human thought of what we call spirit and spontaneity.'[1]

It would be comforting indeed to be *sure* we knew the truth. Man's quest for certainty is his most eager one. To assume that life is simple in the sense that it depends on the known principles of matter and of energy is the easy way to certainty about it. To do so ignores the uncomfortable problems, especially that most obscure of all—how goals arise in living stuff and thus how aspirations of the spirit come to birth. To assume that a thing is simple when it actually is not so, however, is to delude one's self. It certainly is true that circumstances both within the organism and without affect its insights and desires, just as the magnetic needle may be deflected from true pointing to the pole, but we do not count the compass valueless because of this. Instinctive longings of the human spirit may doubtless sometimes be hallucinations, but to maintain that they must therefore always be completely valueless is to hold too dogmatic an opinion about a matter which is far more complex than the confident materialist is willing to admit. Both science and theology are too naïve when they assert that only *their* answers can be true. 'There are more things in heaven and earth, Horatio, than are dreamt of in your philosophy.'

Life is still the final problem and we do well to admit that we are far from a final solution of it and may not even be on the road which will surely take us there. Today is not a time for easy dogmatism in science or religion. If a reasonable pos-

[1] T. H. Huxley, *Methods and Results*, as quoted by E. W. F. Tomlin, *Living and Knowing* (London: Faber and Faber, 1955).

sibility exists, as I believe it does, that what we have called the spirit is life's highest manifestation; that through it we can make a direct, and not simply a logical, contact with the nature of life itself; and that life thus sensed is continuous with a greater source of life from whence it comes—then we should be negligent indeed if we ignored what the spirit has to tell. Through it we may be able to look deeply into the very heart of reality and discover truths that are inaccessible by any other means. Life is a continual seeking after goals, and in following the ones the spirit knows we become explorers in that wide realm which is man's native country and to which he is ever striving to return.

Matter, life, mind and spirit are somehow tied together in intimate association. This reaches its highest level of complexity in the activities that go on within the human brain. Wilder Penfield, a foremost student of the brain, thus concludes a recent paper on this remarkable organ: 'It is obvious that nerve impulse is somehow converted into thought and that thought can be converted into nerve impulse. And yet all this throws no light upon the nature of that strange conversion. Before certain problems the scientist will always stand in awe. Perhaps he may be forced to make another approach—to what was called in old time 'the heart.' However far our successors in these studies may go, it is my belief that the machine will never fully explain the man, nor mechanisms the nature of the spirit.'[1]

[1] Wilder Penfield, *Some Observations on the Functional Organization of the Human Brain*, Proc. Amer. Phil. Soc. 98:297, 1954.

12

Beauty

ONE OF the goals man's spirit has sought through all his history, one of his chief values, is beauty. The strong attraction that it has for him is one of the best evidences that he is indeed a spiritual being. Why should a machine enjoy such a thing, or what reason is there for beauty in a purely material world?

Beauty is a difficult quality to define. Some combinations of form or colour or sound men value, some images or symbols. In nature they are drawn towards such things and seek through art to reproduce them. Not everything seems beautiful by any means. Why should there be this *élite* among the wide range of man's experiences, these few that he prefers to all the rest? Among the sensations reported by the eye and ear, why should there be this aristocratic minority of things that men call beautiful? If he is to make any distinction among objects, it would be natural for him to favour whatever ministers to his survival or his success in life. Instead, however, he oftenest finds beauty in objects or qualities possessing no obvious practical usefulness at all. He hangs paintings on his walls, buys sculpture for his garden, visits art museums, attends concerts and plays, reads poetry, novels, drama and every other form of literature. He encourages artists of many kinds by paying for their services or buying things which they create. More than all this, he tries to produce beauty by the work of his own hands. Often fumbling and blundering but by the hundreds of thousands, in the United States alone, he uses his increasing leisure to sing or play an instrument or write or paint. There is something moving in this almost universal quest for beauty. Men and women who are by no means widely cultivated nor have acquired discriminating taste still strive to satisfy the instinctive longing for the beautiful that rises in their hearts.

The kind of beauty that men seek varies widely and differs among individuals, races and patterns of culture. This is natural for several reasons. Differences exist in the reports that we receive from our sense organs. Inherited variations have been found in ability to discriminate absolute pitch and to distinguish between colours. There are doubtless many lesser differences of this sort among individuals, so that to some degree we live in sensory worlds that are not all alike.

Why people's tastes as to what is beautiful are not similar, however, is due far more to social than to biological causes. The particular culture pattern in which one has been brought up often establishes standards of aesthetic taste to which his own will naturally conform. The music appreciated by the American Indians and by our Pilgrim fathers was widely different and neither would draw an audience to a concert hall today. The music of the Middle East is so unlike our own that at first it is hard to understand it, and that of the Far East is stranger still. The arts of ancient Egypt and of Greece were quite unlike, moulded by differences in bodily character, geography and religion, but we have little difficulty in admiring both of them. Among the national cultures of Europe today there are differences in aesthetic sensibility, as between Englishmen and Italians, some of which may have a racial basis but which stem primarily from diverse cultural backgrounds. There is little doubt that almost any child, placed at birth in almost any culture, would come to regard as beautiful whatever that culture so regarded. Tastes are certainly acquired.

Standards of aesthetic values can be cultivated. Many people, gaining in culture and sophistication, look back with amazement to things they once thought beautiful. Styles in beauty change, too, as in most other values. What was thought lovely in the days of Queen Victoria became ugly and old-fashioned as the new century began, and since that era modes in beauty have become still different. Poets, authors, painters and musicians go in and out of favour. Gray's 'Elegy' was once regarded by almost everybody as the greatest poem ever written. In the USA the New England poets and writers were so venerated that for a long time no one listened to any newer voices. The spirit and objectives of an art may sometimes be altered so radically, however, as in modern painting, music and poetry, that an older generation cannot understand what a younger one is trying to accomplish.

All this has persuaded many that beauty is merely subjective and relative, just what a particular pattern of culture thinks it is, and that no abiding standards exist by which it can be judged, that the acrimonious controversies among critics are so much chatter and that before long today's issues will be forgotten as people become interested in new ones. The scientist, particularly, is apt to be a little smug as he compares the certainties and testabilities of his own material with the doubts and differences and constant changes of opinion he sees in every field of aesthetics. If beauty *exists*, he thinks, it ought to be possible for its devotees to agree on what it is.

In reply, a seeker for the beautiful hastens to point out that what he tries to find, like all other goals of the spirit, is not measurable and therefore that it never can be standardized. A product of the intellect, on the contrary, can be communicated from one person to another so that both have exactly the same data. By exercise of reason it is therefore usually possible to come to an agreement as to what is true. In this way truth can be disentangled from error and progress made. This method reaches its highest development in the sciences and accounts for their prodigious success. Whatever is a product of the spirit, however, is not communicable. No instrument, however exquisitely sensitive, can register one's particular delight when he looks at a great painting or listens to a musical masterpiece. The exalted experience of a mystic is ineffable, but so is everything else that is felt rather than measured. Can you bring the flavour of maple syrup to a man who has not tasted it, or describe a hermit thrush's song to one born deaf? The physical accompaniments of feeling and emotion can be recorded—the changes in breathing and pulse rate and less obvious physiological alterations—but it is impossible to measure the wide diversity and strength of our emotional reaction. We can infer what our comrades are feeling by their behaviour and their words but they can never share these things with us. Each of us is alone with his sensations and his value judgments. It is remarkable not that we sometimes disagree but that we are as harmonious in our tastes as actually we are.

By everyone, however his preferences may differ from his neighbour's, beauty can be recognized. It is rarely associated with a single sensation—a tone, a colour, a word—but with a *pattern*, a Gestalt, an orderly arrangement of impressions. A series of random musical notes means nothing to us but when they are arranged in a particular order and rhythm, as in the opening bars of Beethoven's Fifth Symphony, they stir our emotions. A random spattering of pigments on a canvas means little, but if these have a particular arrangement, a masterpiece emerges. Beauty is orderly, not chaotic. It is an *organized pattern* of sights or sounds or words or images which strikes a chord within us; which vibrates, so to speak, on our particular wavelength. To carry the figure a little further, these vibrations may be in different keys, depending on our aesthetic background, and thus may seem at first to be out of harmony with each other; but each, in its own way, arouses a sense of the beautiful.

This suggests that there may be a relation between an organized pattern that we call beautiful and the organized protoplasmic pattern that each of us is. If our goal patterns, the things our minds and spirits value, are basically the same; in other words, if the patterns in the living stuff of man are at bottom essentially alike, is it not reasonable to suppose that there is a standard for the beautiful which, in a sense, is absolute; that beauty is *whatever is in harmony with the basic goals of human life?*

That such a conclusion is reasonable is suggested by a remarkable fact which has not been given the attention it deserves: that the structures produced by living things are almost always esteemed as beautiful. This may be somewhat too wide a generalization but the fact remains that the bodies of plants and animals, even to their minute parts, arouse in us a sense of beauty. There are differences among them, as among creations of a human artist, but they are all worthy of our admiration. Life is a consummate artist.

Examples of this fact are literally without number. How impoverished our lives would be without the beauty of flowers!

Births, deaths and marriages are graced with them and they brighten almost every corner of life. From the days of Babylon to our own they are one of the chief reasons men plant gardens, and the level of a civilization may almost be measured by the attention that it pays to them. Botanists assure us that the almost infinite variety of the corolla's form and colour, and the fragrance that dwells in it, are mechanisms developed by natural selection to attract insects and ensure a union of the sexual cells. But surely such a profusion of loveliness would not be necessary to entice a bee. Even if floral beauty is simply a biological by-product, it tells us something important about nature's aesthetic creativeness, that decorates a simple process of life with this wealth and over-plus of beauty.

Birds, with their grace and lovely plumage and melodious songs, have delighted men since the dawn of history. From hummingbird to peacock their beauty is one of these bonds that tie men to nature. The universal appeal of birds to man is shown by the thousands of people whose avocation is the study of these remarkable creatures.

And so on down the long list of animals and plants that refresh men's hearts—butterflies, shells, trees, the tiger's flaming stripes and the antelope's grace, tell the same story. All these have come up over the long evolutionary road, their every trait, so the Darwinians maintain, the result of random variation and competitive struggle for survival. Only here and there, as perhaps in sexual selection, is there any indication that this beauty has had practical usefulness. The process has been a strictly utilitarian one, with continuing existence as its only evident cause. Why, then, we ask, has it been interfused with the creation of such a wealth of loveliness? *All* living things, whether successful or marked for extermination, alike are beautiful. The life of earlier ages, long since become extinct, keeps even in its fossil remains the evidence of its former comeliness. Beauty is something associated with *life*. The goals set up in protoplasm lead to its creation.

The universal occurrence of beauty in the bodily forms of animals and plants is evidence, I think, that the direction of life, from its primitive stirrings, is towards qualities that have this spiritual significance. Ugliness, when we find it, is usually in the lifeless, not the living; or if in the living, only when the normal organized control of life breaks down, in galls, tumours, malformations and abnormalities of various sorts which we instinctively abhor. Ugliness is the result of randomness and disorganization. Perhaps the very organizing process itself is one that leads to beauty, as it seems to do in crystals. Beauty's variety and profusion in sound and form and colour, however, are far greater in the products of life than elsewhere. Beauty is of life's very essence. It is one of the permanent and indestructible parts of nature.

> Captains and conquerors leave a little dust,
> And kings a dubious legend of their reign;
> The swords of Caesars, they are less than rust:
> The poet doth remain.[1]

It is therefore not surprising, I believe, that this immanence of beauty in all living things, born in the formless fluid stuff of protoplasm from which we take our rise, should have its influence upon ourselves. The normal human body is a thing of beauty and has inspired artists from the dawn of history. But in man this urge of life towards the beautful is carried far beyond his physical self. It colours all the goals set up within him. He admires the beautiful bodies of plants and animals because he comes from the same living stock as they; but he *feels* this urgency towards beauty, as they do not. He is conscious of it, as of other goals, and this is what sets him to seek out beauty wherever it can be found. He sees it not only in life but in the wide and lifeless universe, admiring it in sea and sky, in mountain and in cloud.

But from his earliest days man has done more than this. Consciously aware of the beautiful and able to direct his course towards it, he was not satisfied merely to admire it where it could be found in nature. He began to create it. The day when first he covered his body with patches of colour that pleased him, or ornamented his clothing or his utensils, or began to sketch his first fumbling representations of animals on bones or the dark walls of his cave, that same day marked his rise from

[1] Sir William Watson, *Lachrymae Musarum*.

beast to man. From the caves of Dordogne, with their marvellously life-like drawings of the beasts he knew, through its slow progress across the millennia, anthropologists can trace the rise of man's artistic sense and his increasing skill in ornament and representation. In chants and by primitive musical instruments he also sought to create the beauty that lies in sound. Song and story began what was to grow into the arts of music and of literature.

Through the years, history records man's constant attempts to create these elusive spiritual values. Persons differ greatly in their skill at such a task. Some men, even among those most sensitive to beauty, have no aptitude to bring it forth. Others, rare and gifted souls, belong to the fellowship of great artistic geniuses—painters and poets and musicians—and with them are a host of lesser men whose talents have brought beauty to mankind and made the world a lovelier place in which to live.

What is it, we may ask, that makes the difference between such men and ordinary folk? More than others they are creators, seekers of high goals, who discover in their hearts such rich and vivid images of beauty that these for them transcend all else. By sympathetic intuition they can perceive the inner core of reality and interpret it to us. As Bergson says: 'Our eye perceives the features of the living being, merely as assembled, not as mutually organized. The intention of life, the simple movement that runs through the lines, that binds them together and gives them significance, escapes it. This intention is just what the artist tries to regain, in placing himself back within the object by a kind of sympathy, in breaking down, by an effort of intuition, the barrier that space puts up between him and his model.'[1] Such men are endowed with skill to convert their images of beauty into masterpieces of artistic deed and word.

These are gifts of the spirit and go by the ancient name of *inspiration*, a breathing of the spirit in man. Its nature is much like those spiritual stirrings which other creative geniuses can feel—in religion, science and the leadership of men. Inspiration, I believe, is the supreme expression of what began as a

[1] Henri Bergson, *Creative Evolution*, Tr. by Arthur Mitchell (New York: Henry Holt and Company, 1911), p. 177.

seeking of the simplest bodily goals. Like other goal-seeking of the spirit it leads men upwards into new realms of being. The seeker after the beautiful follows the same highway as the seeker after the good and the Divine. He enters into the very inner nature of what he sees. The man of aesthetic genius is the supreme example of the organizing, pattern-making power of living stuff. He takes the raw data of experience—sights, sounds, words, thoughts—and transmutes them into something that is in harmony with man's very nature. The portrait painter looks beneath the features of his sitter and paints his inner character. The artist sees a landscape not as it is but as it ought to be to conform to his ideal of beauty. The composer hears in his theme eternal harmonies. Poets try to catch in words the deepest longings of the spirit. These men all are seeking not mere orderliness or pattern but something that conforms to a high goal set up within them, something they feel is in harmony with the greater pattern of the universe. In art, perhaps more vividly than elsewhere, we can see the creativeness of life, the drawing power of something deeply longed for, that ideal perfection which the human spirit craves.

In this way art can minister most richly to the spirit of man, can set before him goals which he should try to gain, ideals in harmony with the very spirit of nature. How, one may ask, does art fulfill this purpose now? Today there seems to be doubt among those who practise it as to what art's function really is. It seems often to have lost its primary concern with beauty. Much of modern music, painting, poetry and literature appears to have a different purpose from that of earlier days. Reality is the watchword now. What is the outside world *really* like, the artist asks. Do not try to gild it or idealize it or make it seem one whit different from what it is. Uncompromising truth is what he is seeking to present, if not about the outer world, at least about himself. No more romantic moonshine about heroes and heroines but the stark facts of what men do and especially of what they feel and think. If all this presents the universe as a place less happy and assured than we used to think it was; if most men are creatures chiefly given over to lust and hopelessness and hate, that may be regrettable but such is the case. We should be less than honest if we did not paint man 'wart and all.' Art's great mission, we are told, is to tell us what we are and what can we expect of life. This is a high purpose, and an artist should never lose his dedication to it. The arts have too long been devoted to fostering deception and illusion. The time has come for stark reality. This we must learn to face as best we can.

Such is perhaps a somewhat exaggerated statement of what the arts attempt to do today but it is certainly the aim of many who now practise them. Surely the truth is great and greatly shall prevail. Reality is something we all need to face. Any attempt by poet or by theologian to deceive us, to entwine us with illusion and false dreams, is reprehensible. We must stand up and face the music, whatever tune it plays.

But to confront us with stark reality is only part, and perhaps the smaller part, of what art ought to do. It is the rankest pessimism to believe that 'there is no new thing under the sun.' If man is ever to improve and change the reality of what he is and does today into a better reality tomorrow, there must be held up before him a picture of what he ought to do and what he might become. This is a task not only for the preacher and the philosopher but for the man of letters, too, reinforced by his colleagues in the other arts. Man is bound to have ideals, for he is a child of life, a goal-seeker born. His ideals make him what he is. To determine what these goals shall be is a task for all of us, but no one has a greater share in this than does the poet, the painter, the composer, and the man of letters. They are pathfinders of the spirit, men who carry the flag forward and plant it where the rest of us can follow. They should never be satisfied merely to make an inventory of today in the belief that this is final but should use the incomparable gift of beauty to lift man's spirits and set high his goals. Life moves towards beauty and it is the artist's function to help us reach it. Beauty speaks to whatever is god-like in a man.

13

Right

IF WE agree that man has a natural sense of what is beautiful and what is not, a similar question at once confronts us: has he also any inborn sense of what is right and what is wrong? Is he *naturally* a moral creature? Often we know he is strongly influenced by what he thinks he *ought* to do. The very presence of the word 'ought' in his vocabulary tells us something about him, for 'ought' is concerned with the relative position, with reference to one another, that his goals assume.

Just as among the many sights and sounds his senses bring him a man regards some as beautiful and some as not, so of the possible courses of action among which he must continually choose, some he regards as right and some as wrong. Some offer values of a most attractive kind—pleasures of sense, satisfaction of ambition, gratification of a hundred other desires. Often one such goal conflicts with another that he feels is higher and more worthy. Which one he chooses will depend on how strong the drawing power of these various goals may be; but which he thinks he *ought* to choose depends on the inner spiritual monitor he calls his *conscience*. Conscience stands outside the contest between goals but, like an umpire, gives the verdict as to which is nearest the individual's ideal of moral action. This verdict is not always enforced, but the significant fact is that so often it is made at all.

Ideas of right, like those of beauty, undoubtedly change with time and place and circumstance. In morals, as well as in aesthetics, there are many who emphatically deny any inner standards by which either of these values can be measured. Both are relative, they say, and to maintain that 'conscience' is native to man's heart (planted there, perhaps, as the still, small voice of God) is to yield to archaic superstition, to follow

K

an outworn creed and to deny the powerful influence of material factors in determining what we think is right.

Few issues about man's nature have been debated as vigorously as this one. In persons sensitive to spiritual things the voice of conscience is so plainly heard that they acknowledge it at once and strive to follow its directive. In others it gives a less clear verdict or is rarely heeded. Many recognize no inner monitor at all. The issue here, as in aesthetics, is whether there is a course of action to which man's higher self naturally turns, deflected from it though he often is; or whether what he thinks is right or wrong is settled wholly by his environment and the culture-pattern under which he grew. The latter is the common viewpoint of psychology today, which sees in man's moral ideas nothing more than the effect of early conditioning. Here is the same sort of controversy as over beauty, and with arguments on both sides much as they were there. The moral issue, however, is more serious since it involves the fabric of society itself.

First, what is it that we call right and wrong? Many attempts to define these terms have been made but none is universally accepted. The Ten Commandments hold the allegiance of many, but others repudiate them as nothing more than another ancient code, with no claim to supernatural authority. The basis for a distinction between right and wrong, they maintain, must be determined by what will best promote satisfactory relations between people in a particular social order, what will yield the greatest happiness and fullness of life. Human beings, if they are to live together, must each give up a little of his individual freedom and submit to a code of behaviour that will protect both his neighbours and himself. These codes will vary greatly, for they are products of the particular conditions under which a culture grew. Murder and theft are usually condemned, but in some primitive societies the taking of human life is allowed or even encouraged under certain circumstances. Sparta did not particularly frown on stealing. Where ancestors are venerated, parents receive honour and respect, but if elderly people are regarded as liabilities they are often disposed of. Observance of specific feasts and holy days is enjoined in many cultures, as are particular rituals and duties. It is no wonder that most anthropologists, faced with this welter of diverse moral codes, conclude that morals indeed are relative and that no firm distinction between right and wrong can be established.

Moral attitudes also change with time. One who reads the history of England for the past three hundred years is struck with the ups and downs of the moral climate—Cromwell and the Puritans, the relapse in Restoration days, the easy times of the Georges, the reaction to a strict morality in Queen Victoria's time and its relaxation at the turn of the century.

The basic quesion is whether man can distinguish right from wrong instinctively, so to speak, through the moral goals set up within him, or whether morally he is indeed an empty organism and gets his codes entirely from the environment in which he lives. That there may be *some* inborn predisposition to certain moral values is suggested by the modern attitude towards the means by which natural selection operates. This process used to be regarded as primarily a competitive struggle for existence between organisms, with survival as the prize to the fittest. Many evolutionists now recognize that the struggle is not only between organisms competing with each other for food and for survival but also between organisms and the factors of their environment. Nature is still red in tooth and claw but the amount of blood she sheds is not as great as men once thought it was.

From these facts some biologists conclude that natural selection often actually puts a premium on mutual aid and co-operation rather than on competition only. The social insects, particularly ants and bees, provide examples of co-operative effort which long have astonished and delighted all who study them. Societies of higher animals, though less closely organized, are effective in promoting the welfare of their members as to food and safety. A colony of nesting birds will rush to repulse an attacker. Individual animals often serve as sentinels or guards. There is frequently co-operation in the complex activities of securing food. In not a few cases animals also come together for what seem like playful activities and sheer sport. The books of Kropotkin, Allee, Montague and others are full of examples of co-operation and mutual aid, up and down the animal kingdom. Even in plants there are instances of association between individuals or between species that are mutually advantageous and which presumably have been developed by natural selection.

It thus seems plausible that a natural propensity for some of the moral values, particularly as they concern the immediate group, may have been instilled into man through a long selective period in which the most extreme and selfish individualists were less likely to survive because they lacked the protection given by their fellows. Something like a social instinct may therefore have been developed which became the forerunner of conscience. It should be remembered, however, that this could have been only rudimentary and that animals and primitive man himself have but the germs of altruism. In a state of nature, selfishness is still the rule and not the exception. It seems most unlikely that from a process that is directed ultimately to survival there should have arisen the high selflessness that distinguishes the moral codes of our great religions and the lives of those who almost everywhere are recognized as saintly.

To lift to such a level the primitive social sense exhibited by some of the higher animals and by early man, something more seems to be needed than mere survival value. Here again, as in the case of beauty and other values of the spirit, man's character apparently undergoes an elevation in its moral goals as he grows further out from barbarism. Just as his appreciation of beauty rises, so does his moral sensitivity and the sort of behaviour he regards as right. If one takes the long view and compare's man's moral codes as the centuries have succeeded each other he will be convinced, I think, that progress has been made, bleak as the moral climate of our time may seem. Among enlightened peoples men are no longer held in slavery, children employed for long hours, petty thieveries punished by death or bears baited for sport. To account for these things on any materialistic theory is very difficult. Such a theory, says President Conant: 'fails to accommodate what I regard as highly significant facts, not facts of science but facts of human history. These are the unselfish ways in which human beings often act with compassion, love, friendliness, self-sacrifice, the desire to mitigate human suffering. In short, it is the problem of 'good,' not 'evil,' that requires some other formulation of human personality than that provided by the usual naturalistic moralist.'[1]

The right may exist in many forms or levels, as does the beautiful, and in each of them the role of man's spiritual monitor is to distinguish the better from the worse. Conscience in this way may be a useful moral guide without setting up standards of conduct in detail. The almost universal judgment of mankind that unselfishness and a willingness to serve one's fellows is right and much to be encouraged suggests that it is a major goal of man's nature. It will express itself in various ways in different situations but is the broad foundation for *all* ethical behaviour. If one loves his neighbour as himself he does not need to be specifically enjoined from murder, theft or covetousness. All the minor details of human behaviour can be spelled out in various ways, but this one admonition underlies them all. To set any behaviour against the background of this basic standard and thus judge whether it is right or not is the task of moral character. Just as man tries to distinguish what is beautiful from what is not, at a given aesthetic level, so he attempts to decide what behaviour conforms to this basic moral goal and what does not, whatever the particular code is under which he lives. To this extent, at least, man seems not to be empty but to come into the world with an in-born feeling that love for his fellows is a high and worthy thing. Only the germ of this has been attained through natural selection. It is not implanted by conditioning, though this may greatly aid—or hinder—its expression. It is, I believe, a natural development of man's questing, aspiring, creative nature, the ultimate goal of his spiritual life. 'The kelson of the creation is love.'

This is opposed to the idea of moral relativism, so often supported by anthropologists; the idea that any set of customs and codes is as valid as any other and that there are no moral absolutes at all. This opinion is by no means universal today, however, even among these scientists. In a recent paper Professor Clyde Kluckhohn emphasizes fundamental human similarities in ethical standards. Says he: 'If, in spite of biological variation

[1] James B. Conant, *Modern Science and Modern Man* (New York: Columbia University Press, 1952), p. 99.

and historical and environmental diversities, we find these congruences, is there not a presumptive likelihood that these moral principles somehow correspond to inevitabilities, given the nature of the human organism and of the human situation? . . . Moral behaviour in specific instances and in all its details must be judged within a wide context *but with reference to principles which are not relative*. . . . What is right for Hindus in 1955 may not be precisely the same as what is right for Americans in 1955 but it will be of the same *kind*. . . . Some needs and motives are so deep and so generic that they are beyond the reach of argument: pan-human morality expresses and supports them. . . . Principles as well as contexts must be taken into account.'[1]

But we can take one further step, I think, in the interpretation of moral conduct as related to our basic theme, goal-seeking as man's essential quality. By the inexorable laws of life, some conduct carries its own reward and some its own penalty. 'Sin' begins at the physiological level. This general problem of pain and evil will be discussed in a later chapter but it is obvious, I think, that whatever interferes with the normal goals of bodily activity must be paid for with the price of pain. Pain is the signal that tells the organism when it is in danger of some bodily ill, a warning that the maintenance of an ideal structural state or physiological balance, the normal physical goals of life, is being threatened.

At the level of the mind, where behaviour is decided, the same principle can be applied. The mental pains of anxiety and regret and the vaguer feelings of unhappiness that cloud so much of life come from a similar failure to reach the goals that are established in the mind and spirit of man. These to some degree are changeable and may be modified by the individual or his environment; but whatever they are, a failure to gain them, as with bodily goals, results in mental pain, in tension, in a sense of being unfulfilled. A host of goals, of bodily and mental desires, are pressing to be satisfied. Some are incompatible with others and the harassed man, drawn this way and that, has trouble to decide what he should do. Often the attraction of

[1] Clyde Kluckhohn, 'Ethical Relativity: sic et non,' *Jour. Philosophy*, 52:663-677, 1955.

one of them is so great that although his mind knows he should not do a particular thing and his conscience tells him it is wrong, he does it anyway. He yields to an overmastering temptation. In other cases the issue may be much less clear, and even a sensitive conscience will have trouble in deciding what the right course is to follow. No simple and infallible guide can make decision; no quantitative comparison is possible, as in science. One yardstick there is, however, that can be applied —whether or not a given course of action will conform to a man's fundamental aspirations and thus yield the greatest satisfaction at last. Many pleasurable goals prove when gained to be short-lived, and some delights that are hoped for never come. The pursuit of happiness is a difficult objective and happiness itself is of many kinds and various degrees. In the long experience of the race, however, men have found that certain kinds of acts, whatever temporary pleasure they may bring, result at last in weariness of spirit, unhappiness or regret and that others yield much satisfaction. Actions can thus be judged by their inner fruits. Man has certain basic values, better developed in some individuals than in others but present in everyone. These are part of his inborn nature, and as he grows in spiritual stature he becomes more sensitive to them. Whatever he does that helps to realize them and thus brings final satisfaction we may call right, and anything that thwarts them and thus leads at last to unhappiness is wrong. This measuring stick is often very difficult to use, for the results of an act may be far-reaching and are often long deferred; but this test provides a foundation for morals which in a sense is absolute, since it is grounded on the character of life's goals and thus of life itself. Conscience is the inner sense of what these goals are, and pain and pleasure result from the degree to which we are successful in attaining them. Thus we may define the right as *whatever is in harmony with the basic goals of human life.*

Essentially this conclusion is well stated from the viewpoint of a psychologist by Professor Hadley Cantril. 'No matter how diverse the social and ethical standards may be that mankind has developed through the ages and in various cultural groups and sub-groups, they are all ultimately attempts to increase the possibility of gaining greater satisfaction in living. And human living—as such—has, I believe, similar psychological aspects due to the similarites we have as human beings.'[1] Jennings carries this idea still further: 'It is the promotion of life, of its fullness and adequacy, that is the right; it is the degradation or destruction of life that is the wrong.'[2]

What we have called the goals of the spirit are still entangled with a host of lower and purely bodily ones in man's animal nature. He is by no means compounded all of sweetness and light. A little lower than the angels he may be, and god-like in many of his attributes, but an animal he still remains. What makes him so exciting and interesting, so significant in the order of nature, is not only that he is the place where matter and spirit meet but that he is the battleground where a higher nature is struggling to emerge from a lower one; where those bodily goals, so necessary for individual survival in his long ascent up the evolutionary pathway, are bing challenged for supremacy by spiritual ones that are beginning to emerge. Two moral codes are fighting in him, the code of the jungle and the code of the good society. The self, as we have seen, is a basic biological fact and therefore *self*ishness is the motive for most animal behaviour—if not selfishness for the individual, as it oftenest is, then selfishness for the family or herd.

As man developed more highly organized societies these have required as a condition for their success that he give up much of the grasping, selfish behaviour so necessary in his earlier course and devote himself to the welfare of his group, since only thus can his own welfare be assured. A society has often been likened to an organism since it is an organized system, not of cells but of individual human beings. The parts of an organism are differentiated, each performing its particular activity but all so inter-related that the welfare of the whole is maintained. In much the same way there is division of labour in a human social organism, each individual's activity being so related to those of others that the society as a whole maintains its health and vigour. If any part or organ of the body fails to

[1] Hadley Cantril, 'Ethical Relativity from the Transactional Point of View,' *Jour. Philosophy*, 52:677-687, 1955.

[2] H. S. Jennings, *The Universe and Life* (New Haven: Yale University Press, 1933), p. 74.

perform its normal function, the goal of the whole will not be reached and there are illness and suffering. Unless by homeostasis the normal state can be again restored, death may ensue. In somewhat the same way, anti-social behaviour by an individual will injure the social group and cause unhappiness and pain not only to himself but to many others. Conduct that violates the goals of society or of the individual is wrong conduct. The concept of self must thus be expanded to include the greater self of society and finally of all mankind. Thou shalt love thy neighbour as thy*self*. It is out of this expanded self that love is born. This new relationship, grown from its simple beginnings in the world of beasts, has helped lift man's goals and aspirations higher than they ever could have reached if his activities were still centred in himself. It is an important step in the expansion of his spiritual life.

The problem of morality, therefore, of what is right and wrong, of what ought to be encouraged or forbidden in man's conduct, must be examined in relation to our fundamental problem of the goals that are set up in his heart. Whether or not morals have a divine sanction involves the great question, so all-important for the validity of a religious belief, of whether these manifestations of the human spirit have any relation to a greater Spirit in the universe itself. It is to this problem that our discussion now must turn.

14

The Divine Spirit

MAN'S highest goals, arising in him as his sensitivity to beauty and to moral values grows, are worthy, I think, to be called manifestations of his spirit. If this is so, we must ask the final question about man's nature, pondered by thoughtful minds from the beginning—does this spirit exist in man alone or is it related to something like it in the universe outside, a Spirit worthy of the name of God? To the materalist and the positivist this question is simply meaningless. Spirit, even in man, is merely a name for certain of his emotions. The idea of anything 'spiritual' in nature, anything belonging to a mysterious supernature, is a ridiculous belief that a sensible generation should have outgrown long ago. Only those things that one can see and touch and measure are real. Intelligent people have no business to believe in fairy tales. 'Good old God' is a pleasant and sometimes helpful myth which man in his ignorance has invented to explain the world. God create man? Nonsense! It was man created God.

In support of this contention it is pointed out that fact after fact that used to be attributed to God's intervention has now been given a rational explanation, and many believe that all phenomena of nature will thus finally be accounted for and nothing left which must be attributed to God. Says Professor J. D. Bernal: 'The role of God in the material world has been reduced stage by stage with the advance of science, so much so that He only survives in the vaguest mathematical form in the minds of older physicists and biologists. In physics He is needed only to explain the creation of a universe which is discovered, as research advances, to be less and less like the one with which we are familiar. In biology He is invoked to account both for the origin of life, and for the purpose of evolution. Now the history of scientific advances has shown us clearly that any appeal to Divine Purpose, or any supernatural agency, to explain any phenomenon, is in fact only a concealed confession of ignorance, and a bar to genuine research.'[1]

There are not many who proclaim an open atheism today for this is still an ugly word and its association with Communism makes it unfashionable, but even among those who are sincerely groping towards a faith, there still are hosts of men who simply cannot imagine that God really *is*. It is this honest doubt, so much less comfortable than the dogmatic certainties of a century ago, that seriously confuses mankind now; but this very doubt holds promise of a deeper understanding of the universe than our grandfathers ever could possess.

Attempts to prove the existence of God by purely rational means have always appealed to men who need a demonstrable basis for their faith. The Church of Rome maintains that 'God can be certainly known by the natural light of human reason from a consideration of the physical universe,' and cites the famous Five Proofs from St. Thomas Aquinas: (1) the proof from motion, that God is the unmoved prime mover; (2) the proof from cosmology, that God is the necessary first cause of things, himself uncaused; (3) the proof from contingency, that God is a being who exists necessarily and not because of something else; (4) the proof from transcendency, that God is transcendent intelligence, truth and good; (5) and the proof from teleology, that God is the designer necessary as an explanation for the orderly universe. Unless one believes in God for other reasons, however, these are hardly likely to persuade him, regardless of their distinguished sponsorship.

The last of the five, however—that the universe, with its magnificent lawful orderliness, is unexplainable save as the work of a great Designer—seems to a philosophical layman the strongest of these arguments. From the days of William Paley men have wondered how such a faultless mechanism as the universe, even if it is nothing *but* a mechanism, could have come to being through chance and randomness alone. If one beholds an intricate machine—an electronic calculator, let us

[1] Quoted by C. A. Coulson, *Science and Christian Belief* (Chapel Hill, N.C.: University of North Carolina Press, 1955), p. 14.

say—it is impossible to conceive of it save as the result of a design framed in a mind. To propose that all the details of it could have arisen without being planned appears preposterous. But this, of course, is essentially what the theory of natural selection did for that superlative mechanism, man. He arose, it says, like all organisms, through the selective accumulation of purely random variations, and there is nothing in the process that suggests a supernatural or a rational purpose behind it, save that whatever survives must conform to the general nature of things. The universe itself is man's creator. Says George Gaylord Simpson, 'Man was certainly not the goal of evolution, which evidently had no goal. He was not planned, in an operation wholly planless. He is not the ultimate in a single constant trend towards higher things, in a history of life with innumerable trends, none of them constant.'[1] And Julian Huxley: 'At first sight the biological sector seems full of purpose. Organisms are built as if purposefully designed, and work as if in purposeful pursuit of a conscious aim. But the truth lies in those two words 'as if.' As the genius of Darwin showed, the purpose is only an apparent one.'[2] Paley's argument never carries quite the weight it used to when one understands clearly Darwin's great hypothesis.

And yet, to many minds, despite these arguments, the universe *makes sense* only on the assumption that there is a creative purpose behind it that is akin to mind. This is no rational proof, of course, but an instinctive judgment, an intuitive feeling that if we knew enough about the origin of things we would find somewhere back of it all a Creator, something more than impersonal, arbitrary law. To be believed in, God must be something *felt* as true rather than proven like a theorem in geometry. Religion would have little power in the hearts of men today unless they experienced a conviction that beyond the world marked out by the physical senses there is one of spirit, accessible to something in man's nature that resembles it and can hold communion with it. To one who

[1] G. G. Simpson, *The Meaning of Evolution* (New Haven: Yale University Press, 1949), p. 292.

[2] Julian Huxley, *Evolution in Action* (New York: Harper & Brothers, 1953), p. 7.

believes that knowledge comes only through intellect, however, and thus in the end from science, such a belief in the validity of inner conviction seems like the veriest nonsense. It is the unwillingness of many men of science to recognize spiritual insight as a guide to truth that is at the bottom of the so-called conflict between science and religion. But if these insights are meaningless, the most characteristic part of man's nature must be left out of account.

Mysticism, derided and condemned by many who pride themselves on being realistic, cannot be tossed aside as meaningless by anyone who seeks a knowledge of what man is. The experiences of the great mystics, as we have seen, are remarkably alike. It is customary today to smile at such men and their burning certainties. Their experiences do not fit into a physical world, but to those who have them they are as real as anything can be. Hosts of men, though not so vividly as these exceptional souls, have felt the power of a spiritual presence in the universe. This to them is such a certainty that it has remade their lives, seizing weak and aimless men and giving them the strength to accomplish goals that were impossible before. Not only do they gain power from this source but an exaltation of spirit that lifts them out of their purely physical selves and into a joyful fellowship with something infinitely higher and worthy of their reverence and worship. Says Canon Raven about such spiritual insights: 'In our moments of awareness we do in fact find that we are caught up and inspired, sustained and encouraged by the power that floods in upon us and the peace that sets us free from fear and self-esteem. A life intense and joyous, an energy which convulses and impels us, a sensitiveness which quickens our insight and deepens our sympathy are the sequel to such moments; and though the revelation fades it does not leave any of us unchanged; none of us can altogether forget, and some are permanently transformed.'[1]

The word 'salvation' has for most of us an antiquated flavour of a religion the chief function of which was to rescue souls from the grasp of sin and hell; but to witness the numberless

[1] Charles E. Raven, *Natural Religion and Christian Theology II* (Cambridge: Cambridge University Press, 1953), p. 57.

cases where a conviction of God's presence has turned the currents of a man's life into a happier and far more satisfying course is to give this old word substance and significance. Just as a sensitivity to beauty may for a little while lift a man up into a mood of spiritual exaltation, a sensitivity to the Divine may do the same, and each provides a basis for some of life's deepest satisfactions.

Why has man sought so long and so earnestly for the Divine? I believe the answer lies again in his very nature as a living thing, a striver after goals that are continually lifted higher as he moves ahead. His love for Divinity is the highest expression of that continual seeking which has marked his course and which shows so clearly in his love for the beautiful and the good. Such goals are the natural expression of man's loftiest values, of what life itself is ultimately bound to seek. To look upon man's search for God as illusion, something imposed upon him from without by his environment or his struggle to survive, or as the natural result of a 'fortuitous concourse of atoms' will never explain it to the final satisfaction of one who has felt the divine presence. 'My soul longeth, yea even fainteth for the courts of the Lord,' says the Psalmist, 'My heart and my flesh crieth out for the living God.' Can this hunger be explained as an illusion brought about by certain curious physiological processes and with no *meaning* as to the existence of that by which alone it can be satisfied? This hunger of the soul, say men of faith, is just as real as the hunger of the body and there must be spiritual food to meet its needs. Are all man's agonies of upward striving, his high goals, his faith in some 'far-off, divine event to which the whole creation moves,' his longing for the hope and consolation that only a loving God can give—are these all empty things and meaningless?

Yes, say the humanists, that is essentially the way it is. Man in his ignorance and fear and weakness desperately needs assurance and to gain it he builds up a 'father image' of a God, simply 'the image of a man, projected, enlarged, upon the empty canvas of the universe.' A God may be necessary but He is not real. Our emotional longing for him means nothing, for it is due finally to physical events in the substance of the brain. Whatever reality there may be to which these are witness is a reality within ourselves and tells nothing about the universe outside, certainly nothing about God. God is man's invention, nothing more.

My colleague Professor Goodenough[1] puts it a little more mildly when he says that what we think are truths about God and religion in general are simply our dreams, myths we have made up to explain the unexplainable, for truth in these deep matters is impossible to reach. This idea at least recognizes man's search for God. Nearer the truth, it seems to me, would be to think of man's relation to God as an active seeking for more than a myth; an *ideal* rather than a dream.

Despite the flat refusal of psychology, and of materialism generally, to accept man's inner conviction of God's existence as valid evidence, this conviction still remains as the solid foundation of all religious belief. If science does not give it countenance, says the man of faith, that is one more proof that science alone is not enough to answer all man's problems. The basic concept of biological organization here presented may make some contribution, albeit a small one, to the task of reconciling these two widely different points of view.

The history of life upon the earth has been a record of advancing complexity, of greater and more intricate organization in living systems. In the higher forms this is far more elaborate than in primitive ones. Some sort of organization, as we have seen, is present in all life, and life may thus almost be defined as organization of a particularly complex kind associated with compounds of carbon, hydrogen, oxygen and nitrogen. The theory of organic evolution pictures life's upward course in the last billion years or so. Occasionally there have been backward steps and the production of simplified and degenerate forms, but almost always the course has been an upward one towards higher levels of complexity. This still continues, but to what final limit we cannot be sure. The remarkable fact, however, is that in the *lifeless* universe, wherever matter is not built into living stuff, the course is not upward but downward, towards a continually greater degree of *dis*-organization. Matter tends to be more chaotic and random. As

[1] Erwin Goodenough, *Toward a Mature Faith* (New York: Prentice-Hall, Inc., 1955).

the physicist puts it, entropy increases. Even a layman can understand why this must inevitably be the case, for particles of matter, whenever they are free to move and at any temperature above absolute zero, are in a continual state of agitation. All other forms of energy tend to degenerate to the simplest one, heat. Such agitation results in constant and random collisions between particles as they dance madly about. However orderly their distribution may at first have been, it is bound to become more and more *dis*orderly. If a hundred red marbles are put in a box and a hundred white ones above them and the box then vigorously shaken, they will become thoroughly mixed and lose whatever orderly arrangement they possessed at first. If the box continues to be shaken there is a chance, a theoretical but almost infinitely tiny one, that they may become sorted out again into layers of white marbles and of red, but the odds against this happening are very great and grow greater as the number of marbles is increased. So is it with the tiny particles of matter, shuffling about and joggling each other like dancers on an overcrowded floor. From this random motion nothing but increasing disorder is to be expected. Not only will any orderly arrangement be destroyed but the energy present will tend to become evenly distributed so that the whole system reaches equilibrium. This is the principle stated in the second law of thermodynamics, the basic description of the physical world. As Schrödinger puts it, this law is 'just the natural tendency of things to approach the chaotic state unless we obviate it.' Life, however, is quite different. It is like the motion picture of a football play run backwards, with the scattered individuals drawing together again into a regular and orderly formation.

This fact confronts us with one of the most difficult questions that science and philosophy have to face. How do living things escape the operation of this law and seem, indeed, to run counter to it? It is possible to explain part of this by recognizing that organisms, as Schrödinger says, can live on 'negative entropy.' Where a supply of energy is available, on its way down towards being degraded into heat, it may be accumulated and stored up, later to start running down again in the work of a machine. The most important case of such accumulation is the synthesis of carbohydrates by green plants through the use of energy captured from sunlight. This is stored up in potential form in food and later released when the food is consumed in the processes of metabolism. It then continues on its downward course. In this respect, living things may be said not to violate the second law but merely to halt its operation locally and temporarily.

In another respect, however, organisms seem to come into more direct opposition to the law. They not only accumulate energy but *build matter into organized system*. A living thing not only draws in relatively simple and random matter from its environment for use as food, thus maintaining the metabolic activities necessary if it is to keep alive, but part of this matter it builds into the specifically patterned forms of its bodily structure. Through these complex and differentiated structures are carried out the co-ordinated processes which make a living thing an organism. Form is the visible expression of this organized activity, and the production of such a form is a very different sort of orderliness from the synthesis of a molecule out of which energy can later be released. The shuffling resulting from heat motion tends not only to disperse the energy of a system evenly but to break down the particular material pattern that it has. A lifeless machine can accumulate energy and liberate it, but an organism in addition builds its own machine. This is an amazing process but it becomes all the more so when we remember that it is done *against* the tendency of matter to disrupt such ordered systems. Life continually opposes this tendency, and as soon as death occurs, the organized body begins to *dis*integrate. Only life seems to oppose this inevitable tendency. In mind this integrative quality is particularly conspicuous. As Eddington says, shuffling can have inorganic causes but sorting into an organized pattern is the prerogative of mind.

The question we must face is *what* this organizing power is. It distinguishes all living things. These differ from the rigid forms of crystals since they are essentially fluid, not fixed. They are 'open systems.' Matter constantly enters and leaves them but their form and character remain the same. To explain the development and maintenance of this organization no satisfactory theory is at hand. The whole course of science has been in the other direction, towards the *analysis* of living things—and lifeless ones, as well—into smaller and smaller parts until it has come to the so-called 'ultimate' particles. Organization, however, is a problem of synthesis and much more difficult. Synthesis implies construction or creation, and what we are really trying to explain is the creative power of life.

The universe thus seems to have within it two opposing tendencies: one downward, mixing, shuffling, leading always towards greater randomness, towards distintegration and formlessness; the other creative, leading upward towards the integration, patterning and regulation of matter. The first is the way of lifelessness, the second the way of life. In our certainty that we have discovered all the basic laws of nature and that whatever unexplained phenomena there still may be will finally be clarified by them as have been so many in the past, we are inclined to shut our eyes on this great fact of biological organization and hopefully to assume that it can somehow be accounted for by the chemistry and physics we now know. This hope, I fear, is destined to be disappointed. No substantial progress towards the solution of the problem has yet been made except to state it with greater clarity.

In this impasse what is the philosopher to do? He should agree with Heisenberg, I think, that if we want to describe mental processes, we shall have to introduce other concepts. Such new concepts will be found, I believe, to concern the facts of biological organization in all its phases—self-regulation, purposiveness, creativity; in short, the distinctive phenomena of life which in these pages we have been discussing. Says Joad, 'I am thus led to postulate the presence in the universe of a dynamic force or principle of change, which enters into association with matter to form living organisms, and is known to the biologist as life, to the psychologist as mind, and to myself as the stream of consciousness which constitutes my being.'[1] How such a principle might work is by no means clear but one may agree it is basic in the universe. I have been bold

[1] C. E. M. Joad, *Philosophical Aspects of Modern Science* (London: Allen and Unwin 1943), p. 271.

enough elsewhere to call it the Principle of Organization and to suggest that it is one of the attributes of God.

This is a startling suggestion which will at once be tossed aside by many. In a subject still so full of doubt and confusion that agreement seems to be impossible, however, it does possess the virtue of drawing in under a single concept a number of diverse phenomena and of gathering them into a single unsolved problem instead of leaving them scattered about.

Spirit, we have assumed, is the highest expression of what has its roots in biological organization, in the goal-seeking of protoplasm. Body, mind and spirit are not separate things but aspects of a single organizing process. They are all qualities of life. Life has moved steadily towards fuller expression until in man there can be seen a glimpse of the possibilities it holds. The human spirit rises from the simplest expressions of goal-seeking in the lower organisms and owes it origin, as they do, to this organizing, creative principle. It is continuous, I suggest, with a Divine Spirit of which it is one manifestation. This organizing capacity, like Promethean fire, seems never to develop spontaneously but to be passed on from generations which stretch back to the origin of life. The assumption of some optimistic biologists that we shall be able to create life is at least debatable. If we do, we shall first need to discover how organization comes about, and the problems to which this is the clue will then be on the road to solution.

Thus we can picture life, man and God as steps in the hierarchical order, manifestations of the same organizing and creative power. This is opposed to the tendency which is evident in lifeless systems. One may call this vitalism if he wishes, but it is a broader concept than that of a psychoid or entelechy resident in a single organism, for it concerns the orderly processes of the entire universe.

One stubborn fact we have on which to build—the *organism*. This has always been an enigma to biologists who have tried to explain it in physical terms. Perhaps we have been approaching the problem of life from the wrong direction. Much, of course, can be learned through the physical sciences about the orderliness of the universe, but in a living organism there are complexities met with nowhere else. Is it not reasonable to suppose that in this living system laws are operative which are unlike those in lifeless matter? Instead of explaining the living organism by physics and chemistry may not the converse also be true, that a fuller knowledge of the organism will contribute richly to our understanding of matter and energy?

If the suggestions that have here been made are sound and if man's spirit, grounded in his biological nature, is indeed like the universal creative spirit and in a sense continuous with it, one should be able to bring together the ideas of religion and of science into a philosophy free from grievous contradictions. This happy consummation will obviously be difficult to achieve. The biologists will be suspicious of any ideas so vague and 'mystical' as these, especially since they involve what seems to be a denial of the validity of one of the basic laws of science. This new concept suggests to him the old abitrary world where God creates and destroys and where the dependability of nature is no more. This is a backward step, he thinks, towards lawlessness again. It should be pointed out, however, that 'laws' of science are not mandatory but simply descriptive and that many have been found invalid in the past.

The man of faith might be expected to welcome a proposal that finds a place for God in a world of science, but he will object to any idea that seems to make God merely a physical principle, part and product of a material system. This would be such a cold and impersonal God that He could never be a vital help to man. Man's spirit, he says, partakes of the Divine and may communicate with God. An essential feature of most religions is that body and spirit are vastly different and that the spirit must shake off its bondage to the flesh before it can find God. Such an objector should be reminded that God's organizing and creative power is only one of his attributes and that it does not preclude a conception of the Divine as rich as can be imagined.

> If God is what everywhere brings form and order out of randomness and finally moulds dead matter into something that gives birth to spirit, He can well be worshipped as Sovereign of the universe, of the lifeless as well as of the living. If man, however humble, shares this Universal

> Spirit, he should be able to make contact with it by that process of communion which through the ages has been known as prayer. If man's spirit is always expressed in human personalities, why may we not expect the greater Spirit, as well, to be manifest as a Person? If purpose is the essential quality of life, and thus of spirit, surely the Universal Spirit is not purposeless but moves towards some great end, some infinite goal. If this Spirit is limitless creative power, it can provide an inexhaustible reservoir of help and strength from which man can draw freely at his need.[1]

So stands the argument that God is man's loftiest goal. The goal-seeking quality of protoplasm can go no higher. Man's spiritual aspirations, born in union with his material body, here reach their final sum. This idea, if it is true, can tell us much about man's nature. God-like in this respect he surely is if he is born of the divine creative spirit. God may be thought of as immanent in all life but only the spirit of man can feel His presence. As Albert Schweitzer says, 'We can find our right place in the Being that envelops us only if we experience in our individual lives the universal life which wills and rules within it. The nature of the living Being without me I can understand only through the living Being which is within me.'[2]

To look at lfe, even in its simplest forms, as the first step in a stairway that leads up to the Divine is both to dignify life as something more significant than a random pattern of proteins and to bring God into a far more intimate contact with man than the God of the theologians sometimes has been. There is a dualism in nature but it is not between mind and body. Rather does it separate the vast and lifeless universe, on the one hand, from living things upon the other. Among these living things stands man, the highest product of creation and the only part of it, we may believe, that shares the creative processes of spirit. Only through man can God be understood. Only in God

[1] E. W. Sinnott, *The Biology of the Spirit* (New York: The Viking Press, 1955), p. 164.

[2] Albert Schweitzer, *My Life and Thought* (London: Allen and Unwin, 1933).

can man be fulfilled. This is the great lesson, I think, that Christianity has to teach. It offers us an incentive to study man's nature as a key to the world of both matter and spirit.

15

Sin, Pain and Evil

AS WE look at man thus risen to his high spiritual estate we can well exclaim, How like a god he is! The amazing triumphs of his intellect open to him material possibilities that are almost limitless. In him life's primitive gropings have been refined to those lofty aspirations by which he seems to hold communion with a greater Spirit in the universe. He is sensitive to life's highest values. He is the summit and the crown of things. No wonder confident humanists predict for him a future in which he will conquer hunger, disease and war, be free from ills and hardships of every kind, control limitless supplies of power so as to live with a minimum of labour, and use the leisure thus acquired to grow in understanding and to find life's deepest satisfactions. If evolution has brought him thus far on the road it surely will not fail to carry him to the end. The one thing necessary is to find the trend of evolutionary progress and to follow where it leads.

The history of this fortunate being from early times till now ought to portray, one would think, a continuous rise in well-being as he progressed along the evolutionary road. Certainly with the phenomenal advances in technology and medicine that man has made in the past hundred years, he should now be safer and happier than in any of the centuries before. His feet should be well planted on the road to the millennium.

It needs no argument, however, to convince anyone today that man has *not* achieved this much-to-be-desired estate. Despite his increased knowledge he is often miserable. Medical skill has conquered many physical diseases but these have been replaced by the more subtle pain of mental ills. Advances in science, far from eliminating war, have made it infinitely more terrible. Standards of human conduct, slowly rising through the centuries, have now too often crashed down to a barbarism where man's cruelty, more vicious than that of savages themselves and refined by exquisite tortures of the mind, is practised by men we thought were civilized. Fear, stark fear, no longer of wild beasts or pestilence or hunger but of our fellow men, is haunting us in the shadow of the unimaginable horrors of total and final war. Crime is more widespread and sinister than at any time in history. Where juvenile delinquency becomes a major problem it is clear that the benign effects hoped for from universal education have not followed it. The millennium seems further off than ever. How *un*like anything divine man now appears, say many. How like a god, indeed! How like a devil, rather.

Voices are heard on many sides today repeating the old, hopeless creed that man is born in sin, a guilty being who never can outgrow his original, fearful taint; that the world is past salvation by the slow progression of knowledge, love and spiritual aspiring and that man's only hope is to throw himself upon God's mercy and pray for miraculous intervention from above.

To one who believes that evolution rather than revolution is nature's way; who is confident that the slow advance of life will continue the task so well begun of drawing our race up from barbarism to the light, such a philosophy of despair seems unworthy of man's great possibilities. If he is being rapidly borne downstream towards a perdition he cannot avoid, the long centuries in which he so laboriously guided his frail craft from the bleak heights of savagery through ever greener lands towards the promise of a rich and fertile valley ahead, have all been wasted effort; blood and tears to no avail and without sense or meaning. This pessimistic philosophy is the natural result of two great wars that have left the world in confusion and despair. It is a reaction, too, against the evolutionary optimism of the nineteenth century and the humanists' too great confidence in man.

In these latter times we have doubtless failed to recognize sufficiently the omnipresence of evil in the world, the evil inherent in mere chance and fate, the evil lurking in men's hearts and the greater evil that sometimes seems to permeate the very nature of things. The problem of evil has long been pondered. One of the most difficult things for religion to understand is how a just and merciful God allows it to exist. Either he is not this kind of God, say many, or else he is not the omnipotent being we have thought. The undeserved suffering of innocence has probably made unbelievers of more people than any other fact. A man who has lost his wife by the long agony of cancer or whose son has been killed by a car in the hands of a drunken driver does not need to be told that there is something sinister in the world. The deep question is how this can be reconciled with the conviction in man's heart that 'somehow good will be the final goal of ill' and that the Divine Spirit is a loving one.

The answer of Christian orthodoxy is that man is inherently wicked, born to evil as the sparks fly upward, and that Adam's fall and his original sin must be paid for by the suffering and sorrow of our race. This doctrine is too unjust and harsh to commend itself to many minds today. It must be recognized, however, that all pain, whether of body or mind, *is* retribution for violating the basic laws of life; if not a violation by Adam, then by ourselves. To be sure, the innocent must suffer with the guilty and the sins of the fathers be visited upon their children. Pain must be expected as the price of sin, whether sins are physiological misdeeds or graver ones. Pain has a necessary value for, like the white stripe painted along the middle of the highway, it warns us when we stray from the right course. Sin, an unfashionable word today, is not so much a theological problem as a vital and practical one. The wages of sin still stand as suffering and death. If we are free to choose our course, we must expect to suffer if the choice is wrong. By his very nature man can distinguish right from wrong and thus avoid much suffering and sorrow.

But what of the pain that guiltless mortals feel, the sufferings of children and of those struck down by chance or accident or disease? What of the undeserved agonies of body and soul that have afflicted righteous men since the time of Job? Can these be attributed to Adam's eating of an apple, however figuratively this action is interpreted? Must we believe that man's nature was so contaminated by this one misdeed that all his seed have been paying for it ever since? Few outside the credal churches subscribe to this today.

Even more difficult to explain on the doctrine of Original Sin is the pain suffered by animals. That they *do* suffer cannot be doubted by anyone who knows them well. Must *they*, too, pay for Adam's transgression, and were they paying for it before man ever was created? Did 'dragons of the prime, that tare each other in their slime' not suffer pain? This difficulty has confronted many who have tried to make sense out of the theory of the Fall. Surely one of the most fantastic of these attempts is that of C. S. Lewis, who assumes a sort of Cosmic Fall which has infected all life with sin and for which the unfortunate beasts through countless millions of years have been paying with their agonies. Pain has a great significance in evolution, but surely it is not *that*!

But if Original Sin does not account for human suffering, what does? Inadequate knowledge, promptly answers the humanist. When man learns to control his environment there is no reason why he *should* suffer. Medicine will some day conquer cancer and the degenerative diseases and perhaps old age itself and man can reach the fullness of his years without bodily ills. Already diphtheria, typhoid fever, smallpox, the deficiency diseases and many others have been almost eliminated. Children live to grow up, and the anguish our forebears suffered from the loss of so many of their little ones is rarely ours. And as to economic problems, the cause of so much distress today, surely our intelligence ought to be able to solve them soon. Already in countries advanced in culture there are produced more food and goods than can readily be consumed, so that the fears of Malthus are not being realized. Sensible economic and population controls are all that is required, and these should not hamper rugged individualists too much. Poverty, the cause of so many of man's ills through all his history, has been largely overcome in the progressive nations and there is no reason why it should not be in all the rest. The poor need not always be with us.

But disease and poverty are not all that troubles people now. Human relationships need radical improvement beyond anything that economic changes can achieve alone. Men hate and fear each other, are suspicious and jealous. They are often selfish, aggressive and intolerant. In short, to use another old-fashioned and unpopular word, they are 'wicked.' It is these traits of man more than economic, social or political ills that threaten society today.

Here the psychologist has an answer to the problem. Poor surroundings and unskilled bringing up thwart and distort the development of children. They acquire inferiority complexes and feelings of guilt which often crop out later in anti-social acts. Frustration leads after a while to aggression. If a person is humiliated or put down in early life, this experience is stored up in his unconscious mind and he attempts to gain a compensation for it later. He is not really sinful or wicked but simply 'ill adjusted,' and for these various ills the psychologist is confident he has a cure. A proper conditioning of the mind by persons who know how to do this—the applied psychologists—will free man of his feelings of inferiority and save him from aggression and other forms of behaviour inimical to society. These emotional maladjustments that disturb human relationships lie at the bottom of much of our pain and misery and are a threat to the very survival of society. Human nature can be modified to almost any pattern if you begin early enough and are skilful enough, says the psychologist, and society reconstructed by adjusting men to fit into it successfully.

All this, if true, is tremendously important. It means that what sets a man right, so that he is happy and society is safe and evil abolished, comes from the *outside*. Science, in the person of the physician, the economist and the psychologist, can make the world over to their hearts' desire. Progress enough has already been made in improving man's environment to give hope that it can help to make his life far richer and more satisfying than it is today.

This solution of the problem is not without its difficulties, however. The vision of a horde of mass-produced, well-adjusted men and women in an antiseptic, painless, sinless world from which all evil has quietly evaporated, even if it could be brought to pass, would attract few normal people. It is too much like the old-fashioned idea of Heaven as a place where there is nothing for an emancipated soul to do but play a harp through all eternity!

An important fact that environmentalists have overlooked, and one that forms the thesis of this book, is that man's character consists primarily of the goals he seeks, the desires, hungers, values and aspirations that are his—and that these are embedded in his nature. They can be modified, elevated or repressed but they are always there and arise from the very quality of life itself. No one who seeks to change mankind can afford to neglect them. These goals are doubtless affected by heredity and can be altered by environmental factors. There are certain basic ones, however, which are much the same in all of us and form the foundation for man's instinctive behaviour; not specific instincts like those in animals, of which there are only a few, but the underlying tendencies and directions of our lives. These often conflict with each other and it is the clash between them that causes much of our unhappiness.

Some of these goals are a direct legacy from our animal ancestry, from days when life was a continual battle for food and shelter and mates, a battle with the inanimate environment and with other animals. A primary fact in biology, as was said earlier, is that life makes individuals, separate selves. It is these selves which are struggling to survive. Selfishness is an ingrained trait in man for it marks an ancient necessity and is required today in some degree to assure his staying alive. It may show itself in many different ways—in ambition, covetousness, cruelty, aggression, greed and hate—and is the cause of most of the sorrow and wrongdoing in the world. Here is the source of that original sin which theologians have talked about so much. Man *is* naturally wicked in one part of his nature. In some individuals and under certain conditions this develops into such a malignancy that one can easily believe such a man to have been 'born in sin' rather than to have acquired it by experience and conditioning. Professor Joad expresses this view vigorously:

> Am I really to believe [he asks] that the passions, the rages, the callous indifference to human suffering, the unbridled lust for domination and display exhibited by the

men of restless energy and dominating will who have fought their way to power during the last forty years, are adequately to be explained as the by-products of a feeling of inferiority engendered by neglect in school? That every guard who has taken delight in the beating and torturing of helpless prisoners in concentration camps was imbued by suppressed feelings of guilt, or that the horrors of, let us say, the Russian Revolution, the Russian purges or the Nazi invasion of and retreat from Russia can be adequately accounted for as the inevitable by-products of revolution and war brought about by a change in the methods which human beings adopt to satisfy their fundamental needs? Is it not obvious that human arrogance and love of power, that human brutality and cruelty, that, in a word, man's inhumanity to man, are responsible for these happenings; obvious, too, that it is precisely these characteristics that have written their melancholy record upon every page of human history?[1]

There also seems to be something malign in the very heart of nature that fights against the good. Many things that cause suffering and are surely evil do not arise from human wickedness nor can they be foreseen or avoided, even by the wisest and most virtuous. They are like the 'acts of God' mentioned in an insurance policy, results of the chance and randomness of so many events. In this sense evil *is* inherent in the world. Chance necessarily opposes man's designs. In his long upward climb he has set his goals ever higher. As a builder lifts the topmost girder into place and then draws up the rest of the framework to it, tier by tier, the force he must continually fight against is the downward pull of gravity which ever threatens to destroy his work. In the construction of its edifice life sets up goal after higher goal, reaching at last towards the aspirations of man's spirit. The force continually opposing this upward climb is the universal tendency towards randomness, disintegration and therefore goallessness. Anything that happens by chance is almost certain to lead to *disorder* and thus to

[1] C. E. M. Joad, *The Recovery of Belief* (London: Faber and Faber, 1952), p. 64.

frustration of our desires. The devil is a long outmoded concept which may have more substance to it than we usually are willing to admit. Perhaps it should be revived and the devil looked upon as the personification of the second law of thermodynamics! Evil, sheer evil, exists in the world and we are wilfully blind if we fail to recognize it.

Certainly some evil is due to our own ignorance of nature and our inability to control her forces. When we have learned to stamp out disease and make life safer and pleasanter for everyone, much pain and misery will be no more. Surely every effort must be made to reach this end and man has already gone a long way towards it. If we let our ignorance reconcile us to an acceptance of nature as it is without bending every effort to conquer evil, sin and suffering, we are not worthy of the gifts that we possess. Despite the optimism of the humanists, however, much of the evil in the world can never be eliminated by the most perfect knowledge and the highest achievements of the sciences. People still will suffer, not only those who deliberately violate the laws of life but the innocent, as well. Nature is inscrutable, as Job in his wisdom concluded long ago. So long as man is a material system, he must expect some pain and sorrow, for these are necessary prices that life pays for its survival and its rise to higher levels. Only through suffering can the highest good be won.

Evils of this kind, however, are simple ones. The deeper sort, the malignant evil that can cast us down from man's high estate and in a very real sense be the perdition of our souls is that which comes from the triumph of our lower, animal desires over our higher human ones. There is no 'problem of evil' in the animal kingdom. The reason, I think, is clear. In man, beside these ancient goals of self that have insured his survival through the ages, there have emerged others which are in direct conflict with them: feelings of love, unselfishness and mercy, recognition of what *ought* to be and thus a sensitivity to right and wrong—in short, man's various spiritual goals. These aspirations are in his heart. He cannot forget them or be sunk so low that he no longer feels their power. We may call them conscience or moral sense or the voice of God or what we will. However their presence is to be explained, they are *there*, the highest expressions of that natural goal-seeking quality all life displays but which in man transcends mere animality and partakes of something far higher. In his heart has always raged a conflict between these two so different goals of his desire. First one will gain dominion over him and then the other. He is in part a spiritual being and a child of God, but also in part a beast and still has appetites which in a beast are innocent enough and partly so in man, but which if followed too far may prevent the attainment of those higher goals which are also part of his nature. 'Sin' and 'wickedness' still deserve a place in our vocabulary. They are not violations of the arbitrary decision of some moralist but describe acts which tend to prevent the realization of our highest goals and thus make beasts of us instead of men. This conflict causes something deeper than physical pain—the anguish of soul that comes from failure to be faithful to the best in us. As a beast, man never can be happy for he feels in himself the call of something higher. As a man, he all too often is unhappy, too, for lower desires continually distract him. This double character is man's distinctive mark and must be recognized by all who seek to understand his nature.

Like a god man is in part but also like a beast. What is godlike about him is that he has the power to make a deliberate choice of which goals he shall follow. He has eaten of the tree of the knowledge of good and evil, however, and if he chooses the evil goal the consequences will bring suffering to him and to his fellows also. Here is the source of the most deadly evil in the world because here a decision can be made that violates the fundamental laws of life and leads to its inexorable penalty. To break a law of health brings physical pain, which is surely evil; but to violate those higher goals which are planted in man's heart will bring him far greater anguish and frustration. Something malign exists in him. Evil can win. 'Truth forever on the scaffold, wrong forever on the throne' is perhaps too grim a view to take of this great struggle, but not wholly so. It was William James, I think, who said that man's life *feels* like a real fight and one of which the outcome is by no means yet decided.

Unless like St. Paul we are 'faithful to the heavenly vision,' the high goal of the spirit set up in our hearts, evil will be triumphant. If psychology can help us follow this vision, we should eagerly seek its aid. If in medicine or science or art or politics there are means to realize it, let us not neglect any possibilities they offer. Religion is the deliberate attempt we make to use every resource of man to reach our highest goal, communion with the Divine Spirit from which we have come and to which, it may be, we shall yet return.

In man the combat rages between lifeless matter, drifting down to chaos, and the living spirit which strives to move towards ever higher goals. The battle is within us, where matter and spirit meet. This is what gives man's life value and significance. By virtue we advance. In sin we are forced to retreat. Pain and sorrow and evil are our casualties. In one small sector of the field we cannot see the master strategy of the universe but we know the side on which we ought to fight and can see which direction the advance is taking. Steadfast courage, faith in our cause and confidence in the final outcome are the weapons we must use to conquer evil.

16

Life and Death

THE VIEW of man's nature we have been defending here is that he is a seeker after goals; goals of many sorts, from those embodied in the architecture of his body and the purposes of his mind to the aspirations by which he tries to mount to the Divine. One advantage that this view possesses is that it accounts for his continual concern about the future. The philosophy of mechanism sets the scientist at work to discover how the human machine has been put together and how it runs. Evolution deals with man's past and the course over which he came to reach his present high estate. Psychology strives to find in the early history of an individual's mind the reasons for what he thinks and feels and does. Science is confident when it deals with accomplished facts, with things that actually happened or are happening now. If the laws it discovers are really universal ones it should be able to predict the future, and this in the physical sciences it does with much success. For living things, however, such prediction is far more difficult. The future, the undiscovered future, is here a subject more for speculation than for certainty.

Man is vitally interested in the future. The past is gone and he cannot alter it, but in the future he hopes for better things. Much of his attention therefore is directed to planning for the days that lie ahead. His life is orientated towards what is to come. If you wish to discover the essential character of a man you will look at his past, to be sure, and will examine his failures and successes and the steps by which he has reached his present state. Much more important, however, is to find *where he is going*, what he wants and plans to do, what the future holds for him. It is his *intentions* that tell us most about him.

M

If a man thus oriented to the future asks science for help he will be told either that the future is already settled by the operation of immutable laws and so one cannot hope to alter it, or—if a certain degree of freedom be admitted—that even then his course will be determined by his environment and there is little he himself can do about it. Nevertheless, poor hoping, planning man, deluded though he may be, persists in worrying about what will happen to him and in striving to direct his fate.

The concept that man by his very nature is a goal-seeker makes possible an understanding of this perennial concern of his with what is to come and provides justification for it. A norm, a goal, a purpose, set up in any mechanism, even a lifeless one, determines what the mechanism will do. It embodies the purpose of its builder, and purpose is always directed towards the future. A recognition of the fact that, built into the protoplasmic stuff of any living organism and especially evident in man, there are a series of goals to be sought, of purposes to be fulfilled, makes it obvious that the primary concern of the organism *must be* for the future. Teleology is the essence of an organism and teleology is pointed ahead. This does not mean that something coming in the future will have an effect on something here and now, but that a pattern in the nerve cells, an image, a goal, a thought will bring to pass a purpose. The bearing of this on the problem of freedom we have already discussed. The significance of freedom is that it is related to the future.

This conclusion from our basic postulate is evident enough and is one of the chief contributions that it makes to an understanding of what man is. By his very nature, man must look ahead. Time's arrow points in that direction and he flies with it. If this is so, he evidently will have a deep interest not only in the immediate future but in the distant one, in the final fate that awaits him and the universe, as well. Since he became a man and realized, as beasts never do, that sometime he must die, this problem of the future after death is one which he has insistently sought to answer.

The final question we must ask about man's nature is this most difficult one of all. Is he an ephemeral being, his presence a temporary episode in the great drama of the universe, or is there in him something of the eternal? Is this life of his, so precious to him and with possibilities so infinite, destined to be snuffed out at last like a guttering candle and to mean no more than that of any animal in the long history of the universe? From man's earliest dawn this question has never ceased to trouble him. Where had his comrades gone whom once he knew but now could find no more? Did they still linger with the body they inhabited or had they disappeared forever in its disruption and decay? Did the *spirit* still survive and haunt the places that it used to know? How universal in the thoughts of early man was this concern with life beyond the grave is shown by the presence in almost all his cultures of complex funerary customs, of carefully prepared graves, mounds, barrows, tumuli and monumental sepulchres. These testify to his belief in a future life and his efforts to minister to its necessities. In all the centuries men have faced this universal problem and have tried to answer it. Every great religion has avowed belief in some sort of immortality and endeavoured to prepare men for it. In Christianity it has occupied a central place. 'In my father's house are many mansions,' said Jesus. 'I go to prepare a place for you, that where I am there ye may be also.' For many souls through the ages, compensation for the misery and evil inherent in the world has been the expectation of eternal bliss when the world is left behind. During the past century this concern with immortality has grown less, partly because of the impenetrable veil which so completely separates life here from any that may exist hereafter, and partly because of religion's greater concern with problems of man's present life on earth. The revived interest in what theologians call eschatology, the second coming of Christ and the establishment of God's kingdom, point to a greater interest in a life to come. Man cannot fail to ponder what his relations are to the Infinite and the Eternal. Is there any contribution to this problem, one finally may ask, which can be offered by our biological concept of man as basically a goal-seeker, an organism whose most distinctive quality is his concern with the future and his continuing search for something higher?

First, as to death itself. Why should it occur? There seems

to be no theoretical necessity for it in the constitution of protoplasm. Many plants, under favourable conditions, presumably could exist without a limit in time. Indeed, a variety propagated by such vegetative means as cutting and grafting is simply the much-divided body of the single individual in which it originated. Of course in all these forms parts of the organisms die, leaves are shed and structures above ground may perish in the winter. In our own bodies cells are continually dying and being replaced by others. This biological turnover is high. In all such cases death is not so much the destruction of protoplasm as the disintegration of an organized system, of an individual. The higher the level of organization, the more certain it is that life will finally come to an end. In a tissue culture, where life is only at the cellular stage, it can be almost indefinitely extended. In plants with a low degree of organization death may be long deferred. All organisms, however, which have attained a high degree of integration and individuality seem surely mortal.

Even here there would appear to be no *theoretical* reason for death. Tissues wear out, of course, and old cells are replaced by new. Waste products and toxins accumulate and finally some vital part of the organism ceases to function. One would think, however, that the almost limitless resources of natural selection would have made indeterminate life possible if it had been useful to a species. The inevitability of death, however, has a sound biological reason—it gets rid of the 'old models,' the existing forms of life, and makes room for the development of new and better ones. There obviously is not space enough on earth for new as well as old, and presumably those types that developed an expeditious way of getting rid of individuals after the span of their reproductive life was over had an advantage because they could evolve more rapidly. At least, this biological explanation of the fact of death makes sense.

Death is never complete, however. An individual passes on its life to its offspring through its reproductive cells. It truly continues to live in those who come after it. Each individual is a bud, so to speak, on the immortal stock of germ plasm which flows on through the ages, whence and whither we do not know. We are like runners in a relay race who pass the truncheon on to those who follow us.

Another sort of immortality is man's through the survival of his influence upon the stream of human life, of the products of his activity in physical objects, works of art and written words, and perhaps most of all by what he does to mould the lives of others and thus extend his existence through theirs. None of us leaves the world exactly as he found it. Often some single act, some small word, may be powerful in the destiny of others. Thus Jeremiah Smith, a plain farmer of Lanesboro, in a brief but telling speech before the deadlocked Constitutional Convention at Boston in 1788, swung by a few votes the support of the pivotal state of Massachusetts to the proposed Federal Constitution and ensured its final ratification. His immortality is in the nation that he helped to found.

But this secondhand sort of immortality, in offspring and in influence, is not what man sought at the dawn of things or seeks most today. There is a feeling of emptiness, of frustration, that comes over one when he thinks that after a long life of creative labour and love and growing knowledge he may come to naught at last, blank emptiness, wiped out and gone. Says Whitman, 'To think that the rivers will flow, and the snow fall, and fruits ripen, and act upon others as upon us now—yet not act upon us! . . . If maggots and rats ended us, then Alarum! for we are betrayed! . . . Do you think I could walk pleasantly and well-suited towards annihilation?'[1] There is something deep within that rebels at a fate like this, that says it is out of harmony with things. 'Our Creator,' Hawthorne remarks, 'would never have made such lovely days, and have given us the deep hearts to enjoy them, above and beyond all thought, unless we were meant to be immortal.'[2]

What makes the thought of mortality so repugnant is not so much that life ends as that it ends while it is still so incomplete. Through the years the poets and philosophers have sought an answer to this problem, this feeling that death is tragic frustration and out of harmony with nature itself. Others than poets have expressed this attitude. In his biography

[1] Walt Whitman, 'To Think of Time.'
[2] Nathaniel Hawthorne, *Mosses from an Old Manse*.

Charles Darwin wrote: 'Believing as I do that man in the distant future will be a far more perfect creature than he now is, it is an intolerable thought that he and all other sentient beings are doomed to complete annihilation after much long-continued slow progress. To those who fully admit the immortality of the human soul, the destruction of our world will not appear so dreadful.'[1]

Why, we may ask, should it not be possible for life to flow on forever, like a river moving to the sea, ever expanding, ever growing greater, without this continual wiping out of what has been gained and this perpetual starting over again? Protoplasm in an individual undoubtedly grows old and tired and is locked more tightly in its material shell. It loses its earlier power of reactivity and regeneration. But even so, could not the evolutionary process have found a way to conserve this life and have achieved progress without destroying the temple of man's body every seventy years? We must be content to believe that nature has good reasons for what she does, but the question always haunts us.

Before this mystery of death is there anything that the science of life can say, any light that can come from the conception that man is a goal-seeking, aspiring organism, not only a creature but a creator, the highest expression of the possibilities latent in protoplasm?

There is little that it can tell us directly, it seems to me. Matter touched with the spark of life acquires the remarkable ability to draw dead matter to itself and make it live; to organize it into self-regulating systems endowed with norms and goals to which all vital activities conform. That which we call death is simply the relaxing of this control, the disintegration of the system and the scattering of the materials that composed it. The machine breaks down, stops and goes to pieces, that is all. Except for the fact that tiny bits of it have been given off from time to time and have made new machines, there is not the slightest hint that the remarkable organizing capacity of life survives when the physical stuff on which it has been operating is once dispersed. What now departs from the body

[1] Charles Darwin, *Life and Letters*.

is not matter, nor is it energy. A body newly dead weighs just as much and consists of the same atoms as in life. It has as much potential energy stored within its molecules. What has gone is a pattern, an arrangement of parts, a system. It is as though a machine broke down because its parts got out of alignment and finally came to a stop. The materialist's argument is simple and to many entirely convincing. The living mechanism wears out, breaks down and stops. Nothing more. That is death; and to think that anything survives beyond the actual material in the body is to imagine that the old family car, when it comes to the end of the road and is broken up at last for scrap, survives as a wandering ghost in some mechanical Elysium.

What answer can the man of faith, the poet and the mystic give to this simple, straightforward conclusion? Many attempts have been made. The Christian doctrine of the Resurrection answers the doubts of many, but there are hosts of others who can never be convinced by its evidence. Communication with the dead by means of mediums and spirit messages has been explored for years and is the delight of charlatans and fakers. Our Societies for Psychical Research have long been gathering evidence in this field, some of which is intriguing if not convincing. To the positivist all this is mere mystical moonshine and beneath the notice of intelligent men. We should remember, however, that strange things happen in the world and should not let our ideological framework become so rigid that every fact has to be squeezed into it or thrown away. Nevertheless, it should be admitted that spiritualism, whatever the credibility of its findings, has given us little understanding of life after death. If the messages it brings from beyond the veil are true, Heaven must be a place filled with trifling activities, and far from the delightful mansions of tradition. Perhaps our receiving instruments, so to speak, are too crude to record the meanings of what happens in any after-life, and that we are like a dog who might learn to talk but could understand little more than the trivialities of his master's existence.

What other possible evidence is there, we may ask, that man's life somehow, somewhere continues after death? There is none that would hold up in any court of law but it is possible, I believe, to frame a hypothesis alternative to that of the materialist which can be explored by those to whom human mortality stark and complete, does not make sense.

The organism that dies, one may readily admit, can well be compared to a machine that breaks down; but what happens to such a machine? Unless it is completely worn out it can be repaired and set going again; not by its own efforts, to be sure, but by a man whose ends the machine will serve. And where did the machine come from in the first place? Whence that design, the disruption of which means death? It was drawn up by a man who built the machine to serve his purposes. The machine itself lacks purpose of any kind.

The living mechanism is a very different thing. It builds itself according to a plan of its own. It repairs itself. More important, in higher forms this plan, this goal, is experienced subjectively as a purpose, the beginning of a mind. The design, the goal in a living organism is spontaneous, not impressed upon it. The designer is the organism itself. Here is the vital difference between a lifeless mechanism and a living one. All purpose. everywhere, comes from life, and the designs of all lifeless machines are born in the creative mind of a living thing. The goals of mind, rising in man to something far above the design of the primitive living system and developing qualities quite unknown in any lifeless mechanisms, are the beginnings of the human spirit, the highest expression of the aspiring, creative quality of life. If the suggestion here presented is correct, this quality is part of the universal organizing, creative power in nature which may be thought of as an attribute of God. This human machine, if we may call it such, is in its various aspects—body, mind and spirit—a part of the creative process of the universe. It is a different *kind* of machine altogether from a purely physical one.

The problem of its immortality therefore brings up questions very different from those involved in a lifeless mechanism. If life partakes of God himself, of the great Organizing Force in nature, it would be strange for this relationship to be dissolved at every death. If man's spirit is part of the Divine, part of the great continuity of creation, this should redeem it from destruction. In some way lifeless matter is drawn into a living thing and there becomes 'alive.' What happens here we do not know. The organizing quality may arise directly from some hitherto unsuspected power in matter or in energy, and if it does, the principle of conservation might save it from dissolution. If this power comes from something that flows in from the creative reservoir in the universe, that enters matter and draws it together into a living system, then we may ask why it should not return again to the source from whence it came.

Immortality is one of those profound problems, like the soul and the freedom of the will, in which many today profess to have lost interest but which, nevertheless, must underlie our whole attitude towards the universe. No seeker for an understanding of man's nature can afford to ignore it. With Tennyson, man continues to ask:

> The wish that of the living whole
>   No life may fail beyond the grave
>   Derives it not from what we have
> The likest God within the soul?

17

Conclusion

AS OUR excursion into the various aspects of man's nature comes thus to an end, what conclusions about him can we draw? How like a god *is* he? What is his significance in the world? Is he worthy of our respect as the highest thing we know, with nothing above him in the universe? Or is he a part of something greater than himself? Can he span the gap between the material and the spiritual and reach towards the Divine? Is he but a chance collocation of atoms, with no more significance than any other machine, or does he share in an unfolding plan of nature as a creative element in the scheme of things? Is he a temporary episode in matter or is the mark of the eternal on him?

The answer a man finally makes to these interrogations will determine not only his personal philosophy but the practical matter of his relations with his fellows and the kind of social order he will build. If men believe that what they are constructing here is a dwelling place for eternal souls, a structure that can become the Kingdom of God on earth, the architecture they adopt will be quite different from that appropriate for an edifice, however splendid, that has no importance save as a temporary tenement for man and is destined to oblivion at last. No small part of the confusion and despair that distract the world today is rooted in a fundamental disagreement as to what man really is.

In Western culture, two answers are most commonly given to the questions we have asked. One is that of Christian orthodoxy. Man, we are told, is a child of God, but a wayward one. He fell from innocence by disobeying God, and for this original sin he and his children still must pay the penalty. He is a guilty creature, helpless to save himself, and only by God's grace can he be redeemed. The world is fundamentally evil, full of sin and suffering, and its chief service is to prepare our souls for eternal life beyond the grave. This is not the answer of all Christians but in a day of pessimism and confusion it is a common one.

The other answer is the confidently optimistic one of humanism. Man by nature is not wicked but good, or at least he can be made good by education and a proper manipulation of his environment. His intelligence, coming to focus in the tremendous achievements of science, is what will save him. Wrongdoing, suffering and other things that we call evil are simply the results of ignorance and can be overcome. Man's future is entirely in his own hands, not in God's, and he has the power to make it glorious.

Neither point of view, however, can compass all of what man is. In these pages an attempt has been made to find a means of reconciling them by bringing man's various qualities and aspects together into a unity. To this end we have tried to find what knowledge can be gained about him from the fact that he is a living being; not merely knowledge about his chemistry and physics, his evolutionary past or his physiological processes and behaviour, but—far more important—how the life that manifests itself in his material body is related to the immaterial parts of him that are so significant but so mysterious in their origin and nature. Our aim has been to fit man into the universe of matter, mind and spirit without the necessity of dismembering him. In a time when the physical sciences so dominate men's thoughts, not only through technology but in their contributions to philosophy, as well, we should remind ourselves that the final problems of man are biological ones. If we learn the laws of life we shall not only know what man is but shall better understand what matter and energy are, as well.

In studying man as a living organism and trying to define his nature in biological terms, we find that among scientists, though there is general agreement as to most facts about him, a wide diversity exists in their interpretation. Three different answers will be given to our questions, depending on whom we ask.

The biologist says that man is composed of some very particular sorts of chemical substances and powered by energy derived from the food he eats. He is in essence a highly complex automatic machine with its own specific character, and similar in many respects to an electronic calculator with which, indeed, he often has been compared. What he does is determined almost entirely by this inner mechanism. To use a crude simile, he is like a mechanical toy, wound up and set to going.

Most psychologists, though equally mechanistic in their ultimate philosophy, have a different explanation for man's behaviour. Internally, they say, all individuals are much the same, relatively neutral and empty, and what one does is almost entirely the result of specific influences from the outside—training, indoctrination and conditioning. Like a television screen or a film in a motion picture camera, he is a sensitive receiver that simply registers a series of external changes.

In neither of these hypotheses is man his own master. He is governed either by inner drive or outer stimulation. The suggestion offered in these pages presents a third alternative which attempts to provide for man's behaviour a foundation in scientific fact but to preserve for him a modicum of freedom from both inner and outer compulsions, and a significant place in nature. It maintains that his behaviour conforms to some inner pattern set up in the living stuff of which he is composed. Evidence for this comes chiefly from the fact that every living thing is an organized, self-regulating system that moves towards a definite end, and when its normal course is disturbed so regulates itself that the end is still attained. This implies something to regulate *to*, a norm or goal to which its activities conform. In bodily growth, the goal is the mature individual. In behaviour, it is the attainment or maintenance of a specific state or condition. Such a goal of behaviour may be regarded as a purpose, a simple psychical act; and from this biological basis the other and more complex mental processes, reaching their highest level in man, have been developed. Man is a complex pattern of such protoplasmic goals. To what they lead, that will he become.

The basic question of man's nature therefore comes down

to this: How are these goals established in his physical being? Are they formed in a purely mechanical fashion in his brain, like a setting on any automatic machine? Or are they moulded by outer forces to which the brain inevitably reacts? Or is it not possible—and this is the alternative that here has been proposed—that there is a *natural, inherent directiveness* in living stuff which, subject to modification by factors inside the body and outside, expresses itself in what a person thinks and does and is? Is not life itself, at least to a degree, *creative*? Does not man thus share in the creativeness of nature, not as a puppet helplessly enmeshed in the rigid certainties of inevitable physical law but as a part, albeit a minor one, of the cosmic creative process itself? If physicists can entertain without dismay the idea that atoms, and perhaps whole suns themselves, are created out of emptiness, why is it not equally conceivable that man, and perhaps every living thing, should be a centre of the generative process? What is created here is not matter but patterns in matter, paths in protoplasm along which energy can move, and thus goals to which growth and behaviour both conform. In the difficult task of reconciling man's freedom and high sense of values with his existence as a material being subject to law, this suggestion merits our serious attention. If cosmogonists can speculate fruitfully, why should not biologists be justified in doing so, if familiar ideas do not explain their facts?

Whether this philosophy of telism, of the protoplasmic goals that guide the activities of living things and thus of man, is a sound one, there will be much difference of opinion. How far it is possible to extrapolate from the biological fact of bodily self-regulation into the sphere of man's mind, his self, his values and his spirit, is still uncertain. These are the most important aspects of his nature. If one is to avoid sheer mysticism he must find some way to account for them which is related to the objective facts of nature. I can find no other means to accomplish this than to regard conscious mind in its various aspects as the subjective side, the inner experience, of this regulatory behaviour of protoplasm. 'The mental life,' said William James, 'is for the purpose of action of a preservative sort. Secondarily and incidentally it does many other things.' Self-regulation is this 'action of a preservative sort' and at higher levels does indeed lead to many other things. If we take this first step I do not see how we can well avoid the next ones, which regard purposes, desires and values as further aspects of this same biological process. The self is thus to be interpreted as the integration of man's various purposive psychical elements.

To look upon the spontaneous, instinctive and emotional side of mental life, the highest level of goal-seeking, as somewhat different from the others, though not distinct from them, and as worthy of the name of *spirit* will evoke still greater disagreement. It seems, nevertheless, to be a continuation of something that begins at the animal level but rises in man to some of the most admirable and god-like of his qualities. Bergson, whose insights here were deep ones, makes a clear distinction between this emotional and intuitive part of mind and its rational counterpart. Man's spiritual side may thus be regarded as his vital self, closer to the main stream of life, in contrast to his rational aspect which has risen from this and is of importance in giving him control over nature and an understanding of the physical universe. There is much to be said for a belief in the significance of the human spirit. A wealth of reality—beauty in its various forms, for example—is inaccessible to reason save in a descriptive and formal sense and can only be *felt*. To make contact with this wide field is the function of the human spirit.

A final assumption—that through his spirit man can communicate with a greater Spirit in the universe—is almost entirely one of faith. God, like beauty, is something that must be experienced, not measured. The fact, however, that the goals of life rise steadily from bodily development through mind to spirit makes it reasonable to believe that this progression may move out far beyond the physical limits of man's life and make contact with a higher and eternal reality. However god-like man may be, if he is a transient creature and there is nothing beyond him, nothing in the universe through which he can complete himself and from which he is able to draw inspiration, he loses significance. If that is man, he is like a candle shining in the darkness, lighting up his little space but destined finally to flicker out and die. But if he is more than this and through his spirit can tap limitless reserves beyond his own being, he shines like an electric lamp whose light is not within itself, as is the candle's, but comes from something beyond. Without God man is unfulfilled.

Perhaps this tragic incompleteness is the fate to which he must reconcile himself. But however cheerfully he may try to do so, something within him still cries out in protest. The very nature of living organisms is to lift the level of their goals, to push up against the downward drag of lifeless matter, to aspire to something higher than they have ever known. Millions of people have been doing so through the generations, and must we believe this all means nothing, leads nowhere and has no significance in the scheme of things? The great gift of the spirit is that it seems to release us from the bonds of the material and the measurable and allow us to catch a glimpse of the unseen glories that lie all around us. Scoff at mysticism as you will, dogmatically assert that it is all illusion, there is, nevertheless, in each of us something that tingles at the touch of a reality beyond the narrow band of wave-lengths to which our limited senses can respond. The positivist who is determined to believe only that of which his physical senses bring him sure report cuts himself off deliberately from what most men have found to be a source both of inspiration and of a deeper contact with reality.

But in all this it is *life* that must give the final answer. Life is what science always tends to underestimate. We need not be vitalists in the narrow sense to realize the immense possibilities that lie in a bit of protoplasm, especially a bit that belongs to the organized living system that is man. To maintain that life is essentially synonymous with mind is to elevate it far above the operation of a machine. To go further and think of the human spirit, too, as something born in protoplasm is to extend life's possibilities almost infinitely. To regard it as a centre of the creative process itself is to make it really god-like. Whatever life may be, even if it is simply the consequence of a particular concourse of protein molecules in an exquisitely complex protoplasmic mechanism, it still remains the source of the highest human qualities. Not only admiration and respect for life but reverence for it, in Albert Schweitzer's famous phrase, should be planted in our hearts.

Our conclusion from all this must be, I think, that man occupies a central place in nature. A scientist once calculated that on an exponential scale man's body is half-way in size between an electron and the universe. This median position in a sense he also holds between the physical and the spiritual realms. His body is made of matter and it runs by energy but out of it there grows a quality that exists in neither and that we recognize as akin to the Divine. The germ of this is present in all life—perhaps even in all matter—but only the germ.

In thus exalting life we must admit that man, its highest expression, is so different from the beasts that he is another sort of being. Although he has clambered up the evolutionary stairway, he has in the process passed a critical point where he became not simply the highest form of life but a different creature from any of those that had preceded him. The account of his origin in the book of Genesis is basically true, for he *is* formed out of the dust of the ground; all things *are* under his dominion; he *has* eaten of the fruit of the tree of knowledge of good and evil, and thus, unlike his predecessors, he is a moral being; and in a real sense he *is* made in the image of his Creator, for he is a child of the spirit and shares in its creative power.

Surely, then, man has reason to think of himself as god-like; a little lower than the angels, it may be, but crowned with glory and honour. But lest in all this he should become over-proud, let him remember that in the immensities of nature he is still a very little thing. No longer does the universe revolve around him or is he is the summit of creation. His home is a small planet circling around a minor star near the edge of an enormous, swirling galaxy. This galaxy, in turn, is one of a billion other island universes scattered through the inconceivably vast depth of space, a cosmos whose size is measured in billions of light-years. It would be strange indeed if life were not present elsewhere in its recesses; perhaps in numberless spheres besides our own, perhaps even a different sort of life from the one we know. It is no one-planet God man worships now, or one whose sole concern has been the creation of our little earth and the salvation of men's souls here alone. Vaster and more cosmic concerns are His. As man comes to realize all this, in the light of what the sciences can now assure him about the universe in which he lives, he must repeat, with even greater awe, the interrogation of the Psalmist, What is man that thou art mindful of him?'

The vast perspective that man now commands and a knowledge of his own physical inconsequence within it may oppress him with a sense of worthlessness. Even the lofty heights that he has reached may seem so insignificant that he quite loses heart. Humility is becoming to him, certainly, but he should not forget that he is a citizen of a far grander universe than his fathers ever could have dreamed of, a Kingdom of a vastly greater God. Time and space are not the absolutes we used to think them, and their mere magnitude should not disquiet us. Man's size is no measure of his worth. His value lies not in his quantity but in his quality. If he has reached a god-like level, even in such an out-of-the-way corner of the cosmos, this cannot be without significance in the greater whole.

A more important reason for humility in man is a realization of how far short he comes of what he might be. He is god-like in a sense, but surely far from this high estate in many ways. Devils there are on earth in human form, as to our sorrow we know well today. Where the power of mind implements the instincts of a brute, there is turned loose in the world something malignant beyond measure. The tragedy of sin and evil haunts man still.

Man's claim to be god-like has but a slight foundation in what he is, and rather lies in what he may become. The great teaching of the doctrine of evolution is that life is not static but can move to higher levels—and to lower ones, as well. We err if we think creation has ended with what man now is. It is his *possibilities* that make us salute him as the climax of creation.

His achievements in the control of nature indeed are great, and unless he chooses to destroy himself, they are bound to increase almost without limit. The physical world and all its forces, so long his foe, at last has been tamed to minister to his will.

More important, however, is the assurance that he can advance in *understanding*. What he now holds to be the truth will certainly change as knowledge steadily increases. His danger lies in growing satisfied with what he thinks is absolute truth so that he abandons the adventure of moving out beyond the old frontiers. Truth exists and is man's final goal. There doubtless are truths worthy to be called absolute and we advance towards a knowledge of what they are, though not yet fully understanding them. Absolute truth is like a mountain still unconquered on whose slopes men are continually climbing upward. We know it is there and occasionally catch a glimpse of the summit through the clouds but still have far to go to reach it. We must not mistake some lower spur for the topmost peak and there remain content. Man who is like a god will never be satisfied if there is further truth to gain.

And more than all, man has immense possibilities for spiritual growth. Of all creatures he alone has had a glimpse of what lies beyond the world of space and time, of things material and measurable. In him life's aspirations have risen past the goals of the body and even of the mind and strive to reach that spiritual core from which the universe takes form. Beauty and truth and love, even imperfectly experienced, give him a vision of what he there can find. These at his best man always seeks to reach. He can gain them and go on to build the Kingdom of Heaven here if he so wills, or he can satisfy himself with lower goals. He is free, and the choice he finally makes will have momentous consequences. This choice man long has recognized. It is well set forth in the words of Pico della Mirandola, an Italian philosopher of the fifteenth century. God, said he, thus speaks to man:

> You alone are limited nowhere, you can take or choose to be whatever you decide to be according to your own will. Not heavenly, not earthly, not mortal, not immortal, did I create you. For you yourself shall be your own master and builder and creator according to your own will and for your own honour. So you are free to sink to the

lowest level of the animal kingdom. But you may also soar into the highest spheres of divinity.[1]

When we admire the god-like qualities in man we should never forget that he has a dual nature, half beast, half god, and that what he will become is still uncertain. Hence come his peril and his promise. His place between the worlds of flesh and spirit makes him the battleground where these two great protagonists are always joined in combat. How the struggle finally will end, on which side he will range himself at last, we do not know. If he chooses the flesh and wills to be a beast, clever but ruthless, the chance for his survival now seems slender, for ruthlessness is armed today with a destructive power it never had before. Only if he rises above his lower nature and follows his highest spiritual promptings will he be able to realize those possibilities that lie within him. Only thus can he complete himself. How god-like he finally becomes will depend on how much like God he tries to be.

Much of what these pages have discussed, one may object, is speculation; speculation with a foundation in biological fact but speculation, nevertheless. It has led to ideas about life and man and nature that cannot be proven true by scientific evidence. Why trouble to erect this unsure scaffolding? Why not content ourselves with that which can be seen and felt and measured, with that of which we can be *sure*?

Such is the position of logical positivism and materialism. It is a defensible philosophy and not a few profess to find it satisfying. Nevertheless, it does not answer many questions man is bound to ask, except to say these questions have no meaning. Answers will certainly be offered, however, by those who demand a more robust philosophy than this to guide their lives. It is vitally important that these answers be not reached through blind credulity or superstition but that they be consistent with scientific knowledge.

The suggestions here presented are an attempt to provide a few such answers that will form a unified and logically harmonious framework of concepts about man and his relations

[1] Pico della Mirandola, in *Moral Principles of Action*, p. 52, ed. by Ruth Nanda Anshen (New York: Harper & Brothers, 1952).

to life and to the universe. They are anchored in science but reach out far beyond it. Upon this framework, I believe, can be constructed a philosophy that will satisfy many who respect the demands of intelligence but also recognize values of the spirit—a philosophy that in the best sense of the word may be called religious.

THE END